Games of Life

Explorations in Ecology, Evolution, and Behaviour

KARL SIGMUND

Institute of Mathematics
University of Vienna

Oxford New York Tokyo
OXFORD UNIVERSITY PRESS
1993

Oxford University Press, Walton Street, Oxford OX2 6DP
Oxford New York Toronto
Delhi Bombay Calcutta Madras Karachi
Kuala Lumpur Singapore Hong Kong Tokyo
Nairobi Dar es Salaam Cape Town
Melbourne Auckland Madrid
and associated companies in
Berlin Ibadan

Oxford is a trade mark of Oxford University Press

Published in the United States
by Oxford University Press Inc., New York

A catalogue record for this book is available from the British Library

Library of Congress Cataloging-in-Publication Data
Sigmund, Karl, 1945–
Games of life: explorations in ecology, evolution, and behaviour/Karl Sigmund.
p. cm.
Includes bibliographical references and index.
1. Life (Biology)—Simulation games. 2. Life (Biology)—Computer
simulation. 3. Game theory. I. Title.
QH313.S585   1993   574' .011—dc20   93-19918
ISBN 0-19-854665-3 (h'bk)
ISBN 0-19-854783-8 (p'bk)

Typeset by Joshua Associates Ltd, Oxford
Printed in Hong Kong

Preface

This book deals with questions of—literally—vital importance, like sex and survival, but it does so in the spirit of a game. In no way is this a novel approach: many recent developments in evolutionary biology have a distinctively playful streak. I have tried to compile some of the highlights in the field, while shunning technicalities, formulae, long-windedness, and Latin names. The book aims primarily at potential or actual students (which includes that interested lay-person so dear to preface-writers). The reader should be ready for games of all kinds, from Cat and Mouse to Computer Tournaments, from Solitary 'Life' to the Only Game in Town, from Hide and Seek to the Battle of the Sexes, from Poker to the Prisoner's Dilemma. There will be lots of games of make believe; but you will not, I hope, find it a trivial pursuit. In fact, I cannot promise entertainment all the way; some arguments by R. A. Fisher or John von Neumann are as intricate as a series of chess moves by Steinitz or Capablanca. But I have tried to avoid all semblance of a method, and kept most of the scholarly routine strictly confined to the (voluminous) references at the end of the book. Ideally, I mean to whet your appetite rather than fill you up. If you wish to read the real thing, or to attend a course on it, or to dig deeper in any other way, I trust that you will find it easy to proceed. Key topics are listed at the start of each chapter, while endnotes provide references or amplify certain points in the text. Each endnote, cued in the text by an asterisk, is preceded by the page number and a brief quote from the sentence to which it refers.

We start with artificial life, and self-replicating machines. This is a topic for computer games like *Life* and other cellular automata. The next chapter is on pursuit games between predators and prey, and on chaotic motion and its role in ecology. This is followed by games of chance and statistical paradoxes arising through randomness in molecular evolution. Some bizarre double games played by chromosomes are treated in the chapter on population genetics. Next come some of the many riddles posed by sex; this is a chapter filled with feedback devices, lotteries, treadmills, and forfeits. The last two chapters apply game theory to animal behaviour. They deal with courtship, ownership, partnership, and brinkmanship, using poker games and computer tournaments.

I have worked and written on some of these questions, but cannot claim to be an expert on cyber-life, ecology, genetics, statistics, strange attractors, parasitology, or animal behaviour: for me, writing this book has also been a gamble. But I had help. It is a pleasure to acknowledge here the debt I owe to many colleagues and friends. Many of my colleagues at the University of Vienna helped me with suggestions and criticisms: I mention in particular Josef Hofbauer, Reinhard Bürger, Immanuel Bomze, Rupert Riedl, Günter Wagner (now at Yale), Peter Schuster (now at Jena), and Martin Nowak (now at Oxford). I have to thank also Jean-Pierre Aubin, John Maynard Smith, Robert May, John Casti, Christian Palmers, Peter Hammerstein, Franjo Weissing, Vincent Jansen, Christoph Krall, Andreas Novak, Minus van Baalen, Ilan Eshel, Ralph Hoekstra, Sergio Rinaldi, Sean Nee, Alan Grafen, and David Haig for discussions, comments, and preprints, and IIASA's librarian Eddy Loeser for his untiring assistance.

Vienna
March 1993
K.S.

Contents

Contents

1

Introduction: Mendel's legacy

'To be, to play . . . Do you know the difference all that well, Chevalier?'
Arthur Schnitzler, *The green cockatoo*

The Spielberg crossings

No formula will give the game away. But I had better confess, right at the outset, that this is going to be a book on mathematics: more precisely, on the mathematics of evolutionary biology.

The links between mathematics and biology are closer than one might think. Gregor Mendel, for instance, the monk with the peas, was a student of mathematics—in my institute, incidentally, at the University of Vienna.* Mendel took many other courses, including some in biology: but biology was not his main subject. He flunked his exam in botany twice, and therefore failed to win a teacher's diploma. Although his teaching earned him much praise, he never was appointed as more than substitute instructor. The son of a small tenant, he entered the Augustine monastery in Brünn (today, Brno in the Czech Republic) at the age of twenty-one. But, as he wrote with disarming candour, he chose to take his vows mainly so as to be able to study in spite of his poverty. Eventually, Mendel was elected abbot and had to spend the rest of his life struggling with Viennese ministries—one of those sad Austrian fates. He no longer had any time for peas, and was obliged to reduce his scientific activities to observing the weather. But at heart, Mendel was a mathematician. 'A young mathematician', as R. A. Fisher described him,* 'whose statistical interest extended to physical and biological sciences.' And since Fisher contributed more than anyone else to statistics as well as to population genetics, we may safely assume that he knew what he was talking about.

It is well-known that Mendel was way ahead of his time: his wonderfully ingenious article *Versuche über Pflanzen-Hybriden* gathered dust for thirty-five years in the obscure proceedings *Verhandlungen des naturforschenden Vereins in Brünn*, until the new century rescued it from oblivion.

The garden where Mendel crossed his peas can still be visited. It is topped by the notorious Spielberg, a picturesque hill crowned by a fortress whose damp dungeons were dreaded throughout the Habsburg Empire. Today, Mendel's monastery is an Institute of Mathematics of the Czech Academy of

Science. A ceaseless stream of traffic passes by. Yet the garden remains: it has not been changed into a parking lot.

In 1924, another gifted young man from Brünn came to study in Vienna: Kurt Gödel, the logician. By 1938, when he left for Princeton, he had shattered the foundations of mathematics. His incompleteness theorem proved that there are arithmetical truths which could, *in principle*, never be proved. This was a very sobering message. Copernicus and Darwin had shown that the position of humans was by no means a privileged one, either in the universe or in the realm of life; since Gödel, it has become clear that even within the sphere of our own creations, our role must remain marginal.

Gödel and Mendel have more in common than just some biographical details. A central role in Gödel's work is played by self-reference. Statements like 'This sentence is false' or 'This sentence kontains three mistekes' refer to themselves. As shown by these examples, such self-reference may provide a twist, a closed loop of contradictions, which makes it impossible to think the statement through to an end. The messages transmitted by biological inheritance also refer to themselves. Indeed, the 'purpose' of a genetic instruction is the production of its like. Again, this is a closed loop, a ceaseless doubling back, a self-reference.

Self-reference and self-reproduction (*self-ref* and *self-rep*, to use Douglas Hofstadter's terms*) are closely related. For instance, it has recently been shown that there is no all-purpose remedy against computer viruses. The proof used Gödel's technique of *self-referential* statements; and computer viruses are, of course, programs for *self-reproduction*. It is no coincidence that the first blueprint for artificial life was devised by a mathematician and logician, John von Neumann (who brought Gödel to Princeton, by the way).

Dry and wet

Most biologists are wary of artificial life, because it is much too abstract for their taste. But sometimes even the most abstract mathematician can come in handy and deliver something solid. After the rediscovery of Mendel's laws, it seemed for a while as if they were contradicting Darwin's theory of evolution, by not allowing enough genetic diversity for natural selection to work on. It took more than thirty years to eliminate this misunderstanding. The mathematical investigations by the geneticists R. A. Fisher, J. B. S. Haldane, and S. Wright contributed most to overcome this crisis; but one of the first substantial advances came from the famous number theorist G. E. Hardy, who flattered himself that he was one of the purest and was prepared to go a long way to escape applications. But he could not leave a fellow cricketer in the lurch, one who happened to be a biologist by trade—and this is why Hardy's name is now to be found in the first chapter of every textbook on genetics, forever linked to a most elementary computation.

The first chapters of many introductions to number theory, on the other hand, start off with a problem from biology: to wit, how fast do rabbits multiply? Fibonacci, the first and for several centuries the only mathematician of renown produced by the Middle Ages, described their population growth by means of the sequence 1, 1, 2, 3, 5, 8, 13, 21, . . . (every term is the sum of its two predecessors). Since then, the properties of this sequence have set many a mathematician's heart aglow.

Hence population dynamics is even older than the kinetics of falling bodies. Why is it, then, that mathematical models are so incomparably more respected in physics than in biology? Has it to do with the fact that if rabbits were really breeding *à la Fibonacci*, they would have filled the whole universe by now? But a stone in free fall cannot accelerate for ever, either: sooner or later it hits ground. Sooner or later, a population meets the limits of growth. All laws are valid only within bounds. Fibonacci was surely aware of this. It cannot be doubted, however, that he was more interested in the arithmetical properties of his sequence than in its applications. There are plenty of bio-mathematicians around who have not reformed a bit.

Nothing sticks like a bad reputation. How often have I been subjected to the tired story of the mathematician hired by an agricultural enterprise to optimize its productivity: 'We shall assume', so his report allegedly started, 'that all cows are spherical.' Even if I manage to stop the story-teller in time, I find myself on the defensive; and I envy those of my colleagues who have turned to physics and are free to toy with one-dimensional Boltzmann equations and stranger fictions still.

Despite all this, no one can deny that mathematical models are playing an ever-increasing role in biology.* As early as the 1950s they were recognized as indispensable (in the wake of spectacular progress in molecular genetics and computer technology). But as soon as one leaves a narrow circle of professionals, one is still apt to encounter a deeply ingrained mistrust: 'You will never be able to grasp Life in those dry formulae', or the like. And it is difficult to argue with this. Indeed, I cannot think of any colleague who has ever displayed an intention to grasp Life. Most of them are wise enough to avoid any attempt at defining it, not even if provoked by the argument that this must obviously be their first and foremost task. They do well to resist. After all, physicists cannot define what time is, or chance, or matter—it seems unfair that no one asks them to. Physicists may, if they like, even dream of a *grand unified theory*, (*GUT* for short): the universal formula upon which everything is based, including life, of course. One may safely predict that such a formula is not going to help with pairs of rabbits. And an engineer constructing a crane finds little help in quantum physics.

So let's stand clear of Deep Thought and the Meaning of Life. More modest questions are tricky enough. How do the population numbers of prey and predators affect each other? Why does the peacock have a tail so long as to hinder its flight? Why do rival deer kill each other so rarely? How old is our youngest common gene? The answers are not obvious at all, even for those who believe that in finding natural selection, we long ago discovered the universal formula of biology.

Some people tend to think that all that is needed are data and a powerful computer. This is way off the mark. One reason is that even the most sophisticated supercomputer is much too small to capture the complexities of, say, the neural network in the human brain or the chemical cycles in a cell; another more exciting reason is that even the simplest models can give rise to calculations which are, in principle, quite unpredictable. This is no different in physics: the oscillations of a double pendulum, the dripping of a water tap, or the tumbling of one of Jupiter's moons are beyond the limits of computability. Again, no one concludes that mathematics has no place in physics.

Obstruct or abstract

Predictions are not the pinnacle of science. They are useful, especially for falsifying theories. However, predicting cannot be a model's *only* purpose. This is not meant as an attempt to falsify Popperianism (I wouldn't know how to begin). But surely the *insights* offered by a model are at least as important as its *predictions*: they help in understanding things by playing with them, just like a child learns much by playing with dolls.

Insights can usually be transmitted without their attendant mathematical apparatus. Again, this is how it is in physics. The theories of Boltzmann, Einstein, Schrödinger, or Hawking have been made accessible to a wide audience, in spite of their extremely mathematical character. Frequently, this is done by means of *thought experiments* which can clarify things although, or even because, they keep their distance from reality.* There is Maxwell's Demon, for instance, sitting behind a sliding trapdoor which he opens only if a molecule comes from the left. There is the astronaut separated from her twin and sent on a dizzying journey through the universe and back. There are those cool-headed observers imperturbably handling meters and stop-watches in an elevator cabin which happens to be in free fall. There is Schrödinger's cat, its fate hanging on a bare thread. And here is a worm on its tortuous way through a black hole.

In theoretical biology, things are not very different. R. A. Fisher speculates about populations having three sexes, J. B. S. Haldane computes for how many nephews he would be prepared to lay down his life, and Richard Dawkins follows the fate of a gene which 'recognizes' the bearers of its copies by their green beards. Such thought experiments are not mere tricks to help convey to the uninitiated some fleeting glimpse of the mysteries of theoretical biology. Rather they are its very essence. The computations will somehow take care of themselves. They will possibly show that the thinking was muddled; or possibly, that one may be on the right track. But in the beginning, there is the model. Calculations will follow.

This is not to imply that mathematics is reduced to mere book-keeping which ought best to be relegated to an appendix. On the contrary, mathematical thinking is the *pre*condition for the thought experiment. It is a relatively secondary task to solve an integral equation or to invert a matrix. What is essential is to take the step into *abstraction*.

Abstraction has a poor reputation: it is deemed pale, unworldly, pointless, and devoid of content. Barren. Mathematics is sometimes *reproached* for being abstract, as if this were an ill-judged step on a slippery slope. But it is precisely this abstraction which lies behind the phenomenal and often quite unlooked-for efficiency of mathematics. The readiness to eliminate from consideration all that seems inessential, to allow for a wider range than that of reality, to compare what actually is with all its possible or even impossible alternatives, this is mathematics' secret of success. This is why mathematics is, in Ian Stewart's terms, the *ultimate in technology transfer*.* The same methods can be applied to problems in astronomy or electrostatics, to the vibrating string or digital data storage, to thermodynamics or the insurance business. The stability of an electrical circuit, a chemical reaction or a mechanical steering gear, for instance, all reduce to the same question.

Toying and tinkering

Admittedly, this may lead to the point where mathematicians no longer notice the diversity behind it all, where they tend to ignore peculiarities and possibly overlook decisive aspects. It is only natural that such a weakness will strike biologists. First of all, because everything strikes them—in contrast to mathematicians, they are trained to observe—and furthermore, because all their experience shows that living nature works without a method. Natural selection constructs without foresight. It improvises, using whatever happens to be around. It lives, so to speak, from hand to mouth. *Evolution is tinkering*, to quote François Jacob.*

Such a biological tinkering is based on the appropriation of available appliances. What serves for thermoregulation is re-adapted for gliding; what was part of the jaw becomes a sound-receiver; guts are used as lungs and fins turn into shovels. Whatever happens to be at hand is made use of. This can lead to strange developments.* The spine serves fish for propulsion, and humans for upright carriage. It is not less astonishing, by the way, when the polynomials used by topologists to classify knots reappear in statistical mechanics, where they specify the behaviour of huge swarms of interacting atoms. Old solutions for new tasks: viewed this way, the differences between biology and mathematics or, if you like, between tinkering and technology transfer, seem not to be so large any longer.

Why, then, do biologists and mathematicians nevertheless tend to think in such different ways? R. A. Fisher claimed that it is due to differences in training.* This does not affect intelligence (which can hardly be trained), but *imagination*. In both curricula, imagination is trained to a remarkable degree, but in totally distinct directions. Biologists become acquainted, right from their first laboratory frog, with innumerable strange beasts of the highest complexity. Mathematicians deal with equations and simple geometrical figures, which seem rather plain in comparison. But their abstract imagination develops thereby. A mathematician learns to think *as if* $x^2 + 1 = 0$ had a solution (so here is a formula, after all—but it will be the last), and *as if* there were no parallel to a straight line; and such thought experiments can become surprisingly relevant. Our universe turns out to be non-Euclidean, and the square root of minus one is a handy tool for electrical engineers. This strengthens, of course, the ingrained readiness of mathematicians to do *as if*. Thought experiments are certainly not their preserve, but it is often useful to let them take a look, and maybe lend a hand. And this is what actually happens.

More and more frequently, biological problems like, for instance, the dynamics of populations, the cost of sex, the development of skin patterns, the emergence of cooperation, the neutral theory of evolution, self-replicating networks, evolutionarily stable strategies, etc. are treated in a spirit more akin to mathematics than to natural science. Occasionally, an animal shows up, as Exhibit A, as it were. Usually, it embodies a counterexample. Computers occur more frequently, but often as abstractions only—like the universal Turing machine, for instance, or some artificial 'biotope' for self-replicating programs.

Such thought experiments are the subject of this book. They will not be covered in any comprehensive way. On the contrary, my selection is highly biased. The chapters are to a large degree independent of each other, which means that they can be read in arbitrary order. And yet, a common thread runs through them all.

Indeed, many of these thought experiments use games and playing, in one way or another. Propagation, for instance, is compared with a *lottery*, in order to explain the advantage of a sexual recombination of genetic material (by buying a hundred tickets, you increase your chances of winning—except if all those tickets carry the same number). Other attempts at explaining sex use a *penny matching game* between parasite and immune system. Similar *pursuit games* describe the complicated population curves in ecology. Self-replicating automata can be animated by means of the computer game *Life*. The mathematical theory of games (evolved from poker and chess and intended to model social or military conflicts) explains all sorts of animal behaviour, as for instance the restrained nature of fighting between stags or the war of the sexes between female and male guppies. The repeated *Prisoner's Dilemma* is used to model the emergence of cooperation in biological societies. *Dice* and *card games* serve to describe molecular evolution. The spreading of a mutant gene in a population corresponds to *drawing tickets* from an urn. And so on. As the physicist J. C. Maxwell said more than a century ago 'we may find illustrations of the highest doctrines of science in games.'*

The notion of play runs like a red thread through the following chapters. In a way, every thought experiment is a play: one plays at *let's pretend* in a realm delimited from reality. Games and plays help to explore the world and teach us how to come to grips with it. Children and cats seem to enjoy it well enough. So, if you are game, let's take this as our cue.

2

Self-replicating automata and artificial life

2

Self-replicating automata and artificial life

Ground rules and the signs of *Life*

... and thus I am inclined to say that a game needs not only rules but also a point.
Wittgenstein, *Philosophical Investigations*

To catch a virus in the net

By now it is confirmed: computer viruses are here to stay. Each strain can be thwarted, for sure, by adequate efforts and means. But it is impossible to find *one* debugging system able to exterminate all strains. This has been proved recently by computer scientist William Dowling, using impeccable techniques of mathematical logic.* It is a certifiably hopeless task to look for an all-purpose vaccine against computer viruses. On the other hand, there is a positive aspect, according to Dowling: people writing detection programs will never be out of business.

The history of network intrusions dates back to the early computers. But it was only in the 1980s, with worldwide access to electronic networks and the proliferation of PCs, that the issue became a major public concern.* A computer virus is a program adept at hiding itself inside other programs and at replicating endlessly. It spreads through infection, using telephone networks or floppy disks as carriers. Frequently, the viruses contain logical bombs—programs meant to damage legitimate files at some preordained time. Some of them have achieved notoriety. One of the best known made the headlines by clogging the BITNET network with season's greetings at Christmas in 1987. Nastier ones are set to strike on Friday the 13th.

The viruses are engineered by people—so far, mostly misguided hackers. But the sinister potentialities are troubling. And even practical jokes which are meant to be harmless can easily get out of hand and cause a lot of disruption. Once released, the virus acquires a life of its own as it starts roaming through the network. Its spread is no longer controlled by its creator. By a strange irony, the age-old human dream of creating life seems to materialize in an obnoxious can of electronic worms. Ah well. Look at what happened to flying, another age-old dream: it turned out to mean sitting in cramped quarters for hours on end.

But happily, artificial life also has some pleasant aspects.

Life can be catching

In the early 1970s, at a time when computer viruses were not yet an all too common plague, there was another type of epidemic causing alarm among computer owners. It used the human brain as intermediate host. The computer game *Life*, first introduced to the world through Martin Gardner's legendary column in *Scientific American*, spread faster than the Spanish flu.* Even those who were immune to it had ample reason to worry. Banks and life insurance companies found that a large amount of their computer time could no longer be accounted for. Many programmers had become addicted to the game, and spent a lot of their working hours on it, always on the alert, in case their supervisor passed by, to push ESC—the escape button—and to wipe the computer screen clear of any treacherous traces of *Life*. It was still worse in many labs and research departments, where the staff did little else, day and night, than fiddle with the infernal game. The aficionados went to the length of publishing a journal of their own: the quarterly newsletter *Lifeline* was entirely devoted to the game.

The person responsible for subverting productivity by inventing *Life* was John Horton Conway, an exceptionally prolific Cambridge mathematician with a bent for oddities and a penchant for puzzles. Conway had invented many games already. He has a knack for it. But *Life* was something special.* *Life* is not a two-person game like chess or checkers; neither is it a one-person game like *patience* or *solitaire*. It is a no-person game. One computer suffices. Even that is not strictly required, in fact, but it helps to follow the game. The role of human participants is reduced to that of onlookers. Apart from watching the game, one has just to decide from which position to start. All the rest proceeds by itself.

The playing field is an unlimited chessboard. Each of its square cells therefore has eight neighbours. The rules are simple. Cells can be empty or occupied. If a cell happens to be empty, it remains empty in the next generation, except if exactly three of its neighbouring cells are occupied: in that case, it will be occupied in the next generation. Conversely, if a cell happens to be occupied already, then it remains occupied whenever two or three of its neighbours are occupied; if not, it becomes empty in the next generation. Instead of empty and occupied, one also says off and on, or dead and alive. This helps to remember the rules. For a birth to occur, there must be three neighbours alive (a classic triangle). For survival, a cell needs two or three neighbours; if it has more, it dies from overcrowding; if it has fewer, from isolation. The cell population changes, in this way, from generation to generation. Its fate is determined, step by step, until eternity. All of its future is contained in the initial condition—there is absolutely no leeway left.

The simple pleasures of *Life*

Now let us watch a few simple games. Never mind if you don't happen to have an infinite chessboard at hand. For the first experiments, a regular chessboard will do, or a sheet of graph paper. We can easily see that if initially only one or two cells are alive, the population goes extinct immediately. But if three cells are alive, their progeny may survive forever. In fact, this happens for two kinds of configurations. With three adjacent cells in a row, the first and last cells die, while the middle cell survives and acquires new neighbours, one to its right and one to its left. So we again have three cells in a row, but the row is turned by 90 degrees. In the next step, the direction flips back, and we get what we started from: we see that this *Blinker* oscillates with period two (Fig. 2.1). If, on the other hand, we start with a triplet which forms a right angle, we get a two-by-two rectangle in the next generation, a stable *Block* which does not change any more: indeed, every live cell has three neighbours and hence remains alive, while every empty cell has at most two living neighbours and therefore stays empty. This is an example of a so-called *still-life*—a pattern which does not change in time.

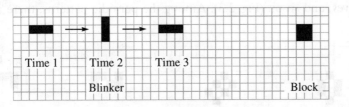

Fig. 2.1 Blinker and Block.

We can also observe patterns which do not alter their shape, or which resume it again and again, but all the time keep creeping over the chessboard. The most important pattern of this kind is the *Glider*. After four generations it looks the same again, but its position has moved one cell diagonally (Fig. 2.2). This motion continues forever, if it does not run into obstacles; the path of the Glider leads straight across the chessboard to infinity.

In these simple examples, the future evolution is easy to predict. In general, however, it is not, which seems strange. After all, the rules specifying the change from one generation to the next are extremely simple. One just has to inspect cell after cell and count how many of its neighbours are occupied. This tells one how the next generation will look. Then, one repeats the whole thing, and so on. The result of one generation step becomes the starting point for the next step.

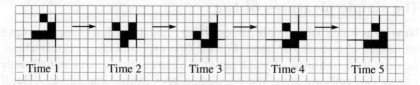

Fig. 2.2 Four steps of a Glider.

Such *recursive procedures* are what a computer likes best. It thrives on closed loops. Its display screen replaces the chessboard to great advantage. The cells (or *pixels*) light up if they are alive. The generations unfold like an abstract cartoon. Oscillators pulsate, Gliders wriggle across the screen, some patterns explode, *débris* assembles into new objects, things grow and scatter, collide and vanish, etc.

Life is a spectator sport.* Onlookers get full value for their time. Experienced observers know lots of familiar still-lifes by sight and name: there are, besides the ever-present Blocks, for instance *Tubs, Boats, Ships, Barges, Canoes, Long Boats* and *Long Barges, Loaves, Snakes, Long Snakes, Ponds, Beehives*, and so on (Fig. 2.3). Among the periodic patterns, there are the

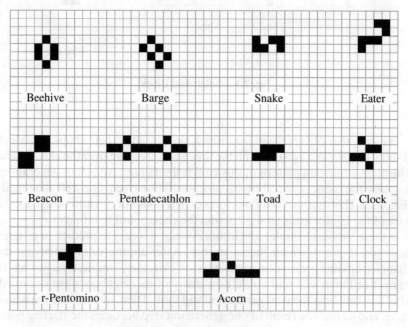

Fig. 2.3 A zoo of *Life*-objects.

Blinkers which we mentioned already, the *Clocks*, the *Pulsars*, and many, many more. Not only Gliders, but also *Spaceships* and *Flotillas* cross the screen.

The variety is quite bewildering. *Life* offers endless vistas of surprise. The more you watch the game, the less you dare predict it. The rules are perfectly known, but the behaviour appears to be completely irregular.

Take, for example, a row of occupied cells. What happens? This depends, of course, on the length of the row. But how? Well, three cells in a row form the familiar Blinker; four cells in a row change into a Beehive; five cells become four Blinkers; six cells in a row do a vanishing act which takes 12 generations; seven cells explode and finally form four Beehives; eight cells ultimately yield four Beehives and four Blocks; nine cells in a row end up as four Blinkers; 10 cells transform into the *Pentadecathlon*, an amazing pattern which repeats itself after 15 time-steps; 11 cells in a row shrink down to a couple of Blinkers; 13 cells do the same; 12 cells, however, develop into two Beehives; 14 and 15 cells in a row fade off, not without a protracted struggle; 16 cells change into a lively *Traffic Light* built out of eight Blinkers; 17 cells turn into four Blocks; 20 cells into two Blocks; 18 and 19 cells dissolve like the Cheshire cat. Nobody has so far found any regularity behind all this. This was of course exactly why Conway enjoyed *Life*.

A cell out for automata

Life is an example of a *cellular automaton*.* Other cellular automata had been studied for decades: one-, two-, or three-dimensional lattices built out of cells. Each cell can be at any given time in one of several possible states. The transitions between states from one time-step to the next depend on the states of the cell and its neighbours. They are determined by well-specified rules— the same rules for all cells and all time. A kind of *micro-causality* governs the evolution of such automata. The states of far-off cells, or of the distant past, exert no influence on what is happening here and now; the next configuration is entirely determined by the present one; the future depends on the past, but only via the present.

Conway had been looking for a cellular automaton whose behaviour was unpredictable, but whose rules were as simple as possible. He experimented a lot before hitting on *Life*. What he had in mind was a model—or, if you like, a metaphor—for a universe whose physics is reduced to a handful of basic rules. In our world, physicists have not yet succeeded in finding such a set of laws: but they are working hard on it, and may well pull it off some day. These laws would then have to account for even the most complex structures in chemistry, biology, and psychology. Some doubt if such a program can ever be worked out. The point of *Life* is to show that even if the physicists were to succeed, the world could nevertheless remain as darkly mysterious as before.

Conway set out to show that some of the most elementary rules can lead to consequences which are far too complex ever to fathom. The local behaviour of *Life* is completely transparent; its global behaviour will never be fully understood. If this statement strikes you as a bit overdone, wait and see.

Conway's first attempts yielded cellular automata whose behaviour was too predictable. His touchstone was the question of whether the system exhibited patterns leading to unbounded growth. If this question was easy to answer, the rules were rejected. The margin was small: obviously, birth and survival had to occur neither too rarely nor too often. But Conway at last hit upon the right recipe: two or three neighbours for survival, and three for a birth. The question about unlimited growth in the *Life*-plane appeared to be really hard.

In fact, Conway found that he could not solve it. This pleased him a lot: he saw that all was well with his game. It definitely deserved additional investment. So Conway put some money on his problem—a fifty-dollar prize promised to the first person to find a pattern growing without bounds, or to prove that no such pattern exists. *Dead or alive*, so to speak. The warrant was published in *Scientific American*, and hundreds of computers started to hum and heat up in the search. Computer time had never been bought at a cheaper rate than for those fifty dollars.

Conway's question, however, could certainly not be solved by just watching the computer. It was not enough to hit upon patterns outgrowing the screen. One had to *prove* that the growth would last forever; that it would not break off after a paltry billion generations or so. The fate of the *r-Pentomino* or of the *Acorn* (Fig. 2.3) shows that such a possibility cannot be dismissed out of hand. These demure little shapes fill the largest display screens with their antics, and quieten down only after thousands of generations. This type of growth is too irregular. One can never know in advance whether it will continue, or stop without warning. What was needed was some sort of *ordered growth* which could be guaranteed to last forever.

Conway gave a hint for a possible solution. People should be looking for *Glider Guns**—periodic gadgets ejecting endless streams of Gliders. A group of students from the Massachusetts Institute of Technology soon managed to devise one: a sophisticated oscillator firing off a Glider every 30 generations (Fig. 2.4). The reward was fairly divided. And *Life* had developed into a cult.

But Conway wanted more. Why had he chosen to call his game *Life*, as a matter of fact? Because it was based on a few simple rules and led to some curious developments? Because it was vaguely similar to the growth of a population, whose welfare depended on being neither too closely packed nor too sparsely spread? Maybe this was indeed what he first had in mind. But *Life* can quickly become an obsession, and so Conway began to take his brainchild literally. No longer was he looking for patterns which grow. Any

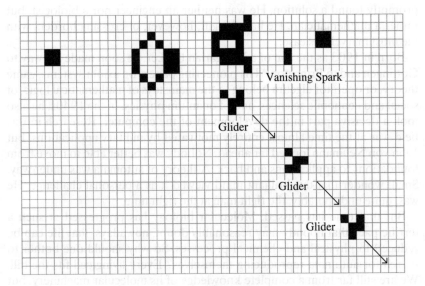

Fig. 2.4 The Glider Gun.

old crystal can grow; so can a dune, or a forest fire. Conway started to look for patterns which were *alive*. Now here was a problem, he felt, that was far too important for public competition. So he solved it himself.

Both the problem and the solution require a bit of explaining. Let us take a long view of what is, after all, a rather long story.

Golem and Co.

Artificial life has always been good copy.* Pygmalion brought his Galatea to life, Rabbi Loew built the Golem, and Dr Frankenstein created Boris Karloff. All these stories are fictions, of course. The real-life inventors and scientists who devoted themselves to the problem left little in the way of historical records, except under the heading of charlatans and fools. There were, to be sure, some extremely skilful clockmakers who constructed mechanical ducks, elephants, musicians, and chess players: but this was not seriously claimed to be life. In fact, as long as the belief in spontaneous generation held sway, the task of creating artificial life could not be properly formulated, let alone understood. Whoever thought of giving it a try headed straight for artificial humans. Today we see that it was only around the middle of our century that the knowledge of biologists and engineers allowed us to pose the question of artificial life in a meaningful way, namely, as the task of constructing an automaton able to reproduce itself. To which task John von Neumann promptly found a solution. He was neither an engineer nor a biologist, but rather a mathematician. He had the gift, however, of crossing borders with the most marvellous ease.

Born in 1903 as the son of a Hungarian banker, he soon made his way to Göttingen and to Princeton.* His pioneering papers on axiom systems, game theory, or operator algebra belong to the mainstays of mathematical logic, of social and economic sciences, and of quantum physics. In the 1940s, he also took a decisive part in the development of the computer. Research in this field was at the time almost entirely motivated by military applications, but John von Neumann looked far ahead. Nobody was better able to move from foundations to applications and back. A theory of automata was under way. Small wonder that this brilliant mind viewed it as a personal challenge. He was, quite undisputably, the right man at the right time.

But that time has passed. Molecular biology has progressed at such a terrific pace that nobody may reasonably doubt, nowadays, that life can be reduced to physical and chemical processes. So why should one bother to search for automata reproducing themselves? Every living cell fits the bill. We are still far from a complete knowledge of its molecular machinery, but the principles are well understood.

By now, scientists are so accustomed to viewing the cell as a chemical plant that they have difficulty realizing that this view has not always been obvious. But there were quite a few, in John von Neumann's days, who held that a mechanistic explanation of life was fundamentally impossible. There had to exist some vital force as yet undiscovered, some *élan vital*, some special *living* state of matter. By an irony of history, the discovery of the double helix in 1953, which effectively led to the demise of all such speculations, took place at the very time when John von Neumann designed his automaton. A confirmation, in some sense, but at the same stroke a devaluation of his efforts. Today, there are still problems in biology which open up like bottomless pits: morphogenesis, for example, or the origin of life. But the replication of genetic information does not belong to this class anymore.

Things change. Some three hundred years before John von Neumann started breathing life into blueprints, philosophers like Descartes or LaMettrie who dared to champion the view that living organisms were mere machines found the question rather embarrassing: why do we never see machines propagate?—What could the good philosophers reply? The machines of their age were rather low-tech. But even today, the question carries some weight. A modern, fully automated car plant, for instance, is a machine producing machines. The complexity of the cars no doubt exceeds the most daring dreams of Descartes' contemporaries by a long chalk. Nevertheless, the car will be considerably less complex than the plant itself. Could this be due to some logical necessity? Must a product always be simpler (in some sense) than its producer? Does the description of a factory need more bits of information, say, than the description of the things which it is able to turn out? A hypothetical law of that sort could be viewed as somewhat similar to the second law of thermodynamics, which states that the disorder of a closed system can never decrease. It would imply that automata can only build simpler automata, and therefore no copies of themselves. If this were true, we would have to throw many familiar assumptions overboard.

John von Neumann did not really believe in such a law. But he wanted to make sure. He had witnessed from the front row quite a few earthquakes shaking science to the core: Heisenberg's uncertainty principle, for example, and Gödel's incompleteness theorem (which states, roughly, that every mathematical theory contains true statements which can never be proved). John von Neumann had probably been the first to understand Gödel's work; he had experienced the shattering loss of certainty which it entails. He knew that we are living on thin ice. To mathematicians, the impossibility of self-reproducing automata, and hence the inconsistency of mechanistic theories of life, would be much easier to stomach than the fact that some statements about whole numbers are fundamentally undecidable.

The haunted warehouse

Abstract biology. From such a point of view, it is a relatively secondary question whether life down here needs some special set of laws, or is kept in being through a supernatural agency, or is a mere delusion of our minds. All John von Neumann wanted to prove was that the mechanistic position was *not unthinkable*. He did not need to actually construct his self-reproducing automaton. It was enough to think it out, in full detail, in order to make sure that there was no room left for any hidden contradiction.

Imagine a kind of robot plodding through a huge warehouse filled with all the parts necessary for its construction: tubes, valves, relays, spools, switches, bolts, nuts, girders, batteries, and what not. The robot fumbles around, gropes for some pieces here and there, joins them and screws them on, and generally plugs, tinkers, and solders until it has finished a robot in its likeness. This new robot gets moving in turn, lumbering through the storage rooms and starting to construct an offspring of its own, and so on.

At first glance, the storage room so conveniently filled with all it takes may look somewhat suspect. But life, life as we know it on our planet, also needs the right stuff at the right place: proteins, fats, sugars, and the like, including highly specialized sources of energy. In an unfavourable environment, life cannot last. So we may just as well ask for the best conditions.

This is not to say that we want to make the task all too easy. All too easy would mean, for example: thousands of identical robots stored in the half-light of a huge hangar. We switch one of them on. It shuffles around until it stumbles into another robot, and switches it on. Two robots are now ponderously groping around for other robots to switch on; soon, a platoon of robots is awake. If we define a robot as alive if it is switched on, then we can claim that it propagates.

This solution is of course without the slightest interest. Something reproduces indeed—namely, the state of being switched on—but we will not be prepared to say that the automaton produces copies of its kind. These copies, after all, had been loitering around already, only waiting for a push to get going. John von Neumann speaks of *trivial self-reproduction* in such a case. It is like the growth of a crystal in a solution. The crystal grows because the right atoms join up; if it breaks, each part grows on its own. This is self-reproduction of a sort, but hardly deserves to be labelled *life*.

The catch is not, by the way, that the robots are fully finished to the last bolt. We could just as well imagine that they consisted of an upper and a lower part, stored in different rooms. If a robot has only to place an upper part upon a lower part in order to obtain a copy of itself, we have trivial self-reproduction again. And this would not be any different if the robot consisted of ten, or of ten thousand parts just needing to be assembled in the right way.

The core of the problem of self-reproduction concerns *information*. An automaton needs to be told what to do. If the program of the automaton is already contained in some part, or parts, stored away in the warehouse and all set to be put in their proper slots, then the real question has not been addressed. For then, the information is not copied at all: it had been around from the beginning, neatly filed and kept on shelves. Living organisms reproduce, as we know, in an altogether different way. They contain a coded program which instructs them to make copies of themselves, including copies of the program, too.

We can see that the task which John von Neumann set himself was not so completely abstract after all. He had couched it in terms which show that he was fully abreast of the state of the art and ready to use the most recent knowledge of organisms and automata. Computing machines had been around for a while; but general-purpose computers with stored programs were brand-new. And it had been discovered only a short while before that the genetic information of a cell was contained in a tiny fraction of its nucleus, in the form of DNA molecules, God knew how.

John von Neumann decided therefore that his automaton would have to consist of two parts: one flexible *construction unit* able to build things out of the elements stored in the warehouse, given the proper specifications; and one *instruction unit* telling the machine how to construct a copy of itself. This corresponded to the duality between computer and program; and also, of course, to the duality between cell and genome.

The incredible shrinking men

Now here a paradox seems to come up. The instruction has to contain, in some way or other, a full description of the automaton: not just of the construction unit, but also of the instruction itself. The copy, after all, will have to know what to do: namely to copy itself, in its turn. Hence the description has to contain a description of itself; and therefore a description of the description, and so on. This yields an *infinite regress* looking desperately like those self-referential statements used by Gödel to shatter the fond dreams of many a mathematician. It also brings to mind an assertion of Paracelsus: according to him, the male sperm contains tiny homunculi which only have to grow for a bit before taking up their place in the world.* Quite apart from being a deplorable example of male chauvinism, this idea strikes us today as utterly muddle-headed: for the tiny little fellows would have to have some sperm, containing tiny little fellows in turn, and so on. A few generations of such miniaturizing, and we are well below the size of electrons. Old Paracelsus did not have to worry: he was probably not an atomist himself, and by his time, the world was not supposed to last for more than a few thousand years anyway. But nowadays, the preformationist's view suggests an infinite sequence of shrinking men. The notion of an infinite sequence of nested instructions seems just as preposterous. The robots ought to be finite, one feels. Makes them more human, too.

In fact, the solution to the apparent paradox is ready to hand. The instruction does *not* need to contain itself. All it has to do is to ensure that it gets copied. It suffices, therefore, that the construction unit includes a machine for copying the program. When the next construction unit is assembled, the automaton hands it down a copy of its own instructions. No vicious circle left.

Simple as it is, this clarifies an essential point. The program has to be used in two different ways: it gets translated *and* copied. In one role, it is in command, causing a sequence of activities; it manipulates the machine. In the other role, it is manipulated by the machine; it constitutes the passive object of the copying unit. In other words, the *uninterpreted* program is just raw data for the duplication; while if it is *interpreted*, it directs the duplication.

When molecular biologists uncovered the actual mechanism of replication, this basic distinction was confirmed. The genetic information, which is stored in the DNA molecule, is encoded by a sequence of nucleotides. Its message gets translated into sequences of amino acids. The resulting *enzymes* are the machine tools of the construction unit. Some of them are *replicases*, corresponding to the copying unit. Their job is the duplication of the genetic information. The genetic code, which runs the cell by translating triplets of nucleotides into amino acids, has nothing to do with this process of duplication. A copying machine need not 'understand' the messages it replicates.

(A *virus*, incidentally, would correspond in this context to an instruction which replicates itself, not by specifying the construction of a copying unit of its own, but by commandeering the unit built by another program. And still more incidentally, the reason why there will never be a universal system able to detect all *computer viruses* is based on the fact that a program is not only something to handle input with, but can be used as an input in its own right—which brings us back to the dual role of an instruction: interpreted and uninterpreted.)

The Turing chameleon

John von Neumann had no intention of imitating the chemical processes in a living cell. His *kinematic* automaton used mechanical and electrical means. For managing this robot, one needed a computer, to be sure. This computer would have to tackle a mind-boggling variety of tasks. John von Neumann had no wish to go into the gory details. He decided to use a *universal computer*. It made things so much simpler for him.

The universal computer had been devised by Alan Turing* in the mid-1930s, at a time when there were hardly any real computers around—just a few dull desk calculators, glorified cash registers, and unwieldy punch-card machines. The universal computer is something different. It will never be improved upon. It can compute *anything* that is computable. The most daunting task performed by the best computer money can buy can also be solved by the universal computer. It will need a very, very long time for the job. But it can do whatever another computer can do.

There is no magic behind it. Every computation needs a program. If the computer XY can perform this program, the universal computer can tackle it too, as soon as it is provided with a coded description of the way XY works. The universal computer simply mimics XY. In a sense, the *instruction* is all that really matters.*

The behaviour of any machine can be formally specified by an abstract set of transition rules. Conversely, any such set of rules can be viewed as a potential machine. General purpose computers are what Chris Langton, a young American who became a leading figure on artificial life before even completing his Ph.D., called *second-order* machines:* they have no behaviour of their own, but must be given a program, the formal specification of a machine which they then 'emulate'. A Turing machine is simply a general purpose computer, taken literally. If you remember Woody Allen's film *Zelig*, the story of a human chameleon—well, Turing's computer is the perfect machine chameleon. It can imitate anything; by the same token, it has no personality of its own.

The most common type of Turing machine uses a strip of paper tape marked into squares. There are digits written on these squares—zeros or ones. All information can be coded by such a sequence of digits. The machine hovers over a square and reads off the digit; it can alter this digit; it can alter its own *inner state* (one of finitely many); it can move by one square, back or forth; and it can halt. That's it.

Everything has its price. The fact that the machine is utterly versatile, ready to grapple with any conceivable algorithm, makes it clumsy beyond endurance. Even the simplest little calculation can take billions of steps. This is the reason why the universal computer has never been used for any practical purpose. But it is priceless as an abstraction: as the ultimate computer for thought experiments. It helped Turing with the solution of the famous *decision* problem, which is of the same calibre as Gödel's incompleteness result, and in fact closely related to it. (We will return to it in a moment.)

John von Neumann was deeply impressed by Turing's work. He viewed his own self-reproducing automaton as a thought experiment, too. To use for its steering unit a computer which was nothing less than universal came quite naturally, under the circumstances. It made the task easier, not harder.

There still remained the small matter of the *external memory* which a Turing machine has to have. It must be infinite, in principle: indeed, since the machine is expected to handle arbitrarily complex problems, it has to be capable of storing arbitrarily large amounts of data. But the automaton's computer has only one aim in life: to make the thing reproduce. It needs no spare capacity in its brain. A huge but finite memory will therefore do. John von Neumann constructed this memory as a very, very long ladder, with some rungs missing and some not. The sequence of rungs coded the information. One just had to make sure that the warehouse held plenty of spare parts for constructing ladders.

The looking-glass cell

It is remarkable that both for computers and for automata, the deepest problems were solved first: universality and self-reproduction.* Neither of the two topics has any practical relevance, by the way. But it is well worth noting that both Alan Turing and John von Neumann were eminently capable of delivering very solid contributions to immediate purpose, if they set their minds to it. During the Second World War, they played, each on his side of the Atlantic, decisive roles in the silently raging 'Battle of the Boffins' which did so much for allied victory. Alan Turing was one of the leading scientists engaged in project Ultra, which cracked the German code. John von Neumann headed a theoretical division of the Manhattan project which produced the first atomic bomb. Turing was possibly sorry for having had to leave the ivory tower; John von Neumann was certainly not. He stood with both legs firmly on the ground and liked people to know it.

So when he felt dissatisfied with his kinematic model because it was *not abstract enough* for his taste, he had substantial reasons. John von Neumann was no woolly-headed dreamer shirking reality. But he was afraid he might have left some loopholes in his argument. Maybe that he had solved only part of the task in devising the robot-building robot with its ladder-memory. After all, here was a problem in *logic*: electrical or mechanical gadgets should have no part in its solution. The kinematic robot was too much from this world. It was based on physical laws which were not totally understood. An engineer would have no complaints, but a philosopher might still harbour the suspicion that somewhere, concealed behind these laws, some unknown vital force or supernatural agency was at work.

In order to dismiss such doubts, one needs a world without any hidden nooks where vitalism might seek refuge, a world transparent down to the very bottom. The thought experiment of the self-reproducing automaton can only clinch the matter if it is performed in such a limpid universe. Now *our* universe is not limpid (thank God). So we must use in its place a fictitious world devoid of any conceivable secret. We have first to create it, and then to people it with self-replicating machines.

This is how cellular automata entered the scene. They were proposed by Stanislas Ulam, a friend of John von Neumann from his Los Alamos days.* Ulam had been quick to appreciate the tremendous opportunities offered by computers to applied mathematics. He was one of the first to use them in a big way, not just for crunching numbers, but for stimulating imagination. Ulam had gained enough experience with cellular automata* to know that waiting here were literally *worlds* to discover—alternative worlds of amazing diversity. Worlds where one need never fear that hidden mysteries could upset the game. The basic rules, the laws of nature in such a universe, are always perfectly known, right from the beginning—they are explicitly laid down in the description of the cellular automaton. The programmer is the only God in residence there.

So John von Neumann created a two-dimensional cellular automaton, whose squares could assume some twenty-nine different states, one of them being earmarked as the 'empty' state. He laid down the recursion rules governing the transitions from one generation to the next, thereby specifying all that could possibly happen. And he devised a pattern which reproduced itself. It consisted of some two hundred thousand cells, and was again composed of two parts: (1) a construction unit including a universal computer, and (2) an instruction programming the replication. And here is how it goes: the construction unit grows an arm extending into an empty region of the two-dimensional board, gropes from one cell to the next, and constructs a replica of itself; then it hands over a copy of the instruction and withdraws the arm. Mission completed.

John von Neumann did not live to finish his book* on self-propagating automata. He died of cancer in a military hospital, attended by security men to ensure that he would not disclose any secrets of state in his delirium. His notes on self-reproduction did not count as secrets of state: they were completed and published by one of his collaborators, Arthur Burks.

The loop tightens

The details of John von Neumann's designs were less important than his general line of approach. There are, after all, many ways in which to attack the problem of artificial life. The earliest, dating back at least to the Greeks, had been to copy the form of living things as perfectly as possible: *true to life*, like the waxworks at Madame Tussaud's. Later, the challenge was to make things move. The most famous example is Vaucanson's artificial duck, built towards the middle of the eighteenth century and unfortunately lost today, a marvel of mechanical ingenuity able to drink, eat, quack, and splash about. Its wings contained over four hundred articulated pieces. Two hundred years later, the interest had shifted from animation to replication. Gone was the quack of the duck; gone were the flesh and bones and all that is called, in the jargon of today's computer biology, the *wetware*. It's all reduced to coded blueprints nowadays. And it can't be denied that John von Neumann's approach has been, ever since the early 1960s, the dominant if not the only approach to artificial life.

John von Neumann's design was greatly simplified by the American E. F. Codd. Codd's cellular automaton was much more suitable for reaching out and transmitting patterns.* A still simpler example was devised by Chris Langton.* Hitherto, there had been a tendency to define self-reproducing units as *non-trivial* if they contained universal computers. Since our own cells can hardly qualify on that account, such a definition was a bit harsh for wetware like us. Langton required, instead, that the information contained in the unit should both be *translated* and *transcribed*. This sounds familiar, and reasonable too. Langton constructed a very small automaton propagating itself. It fits into a field of 15 x 10 cells and consists essentially of a data path cycling along a square loop. This loop is surrounded by a protective sheath (Fig. 2.5). At its south-east corner, the data path spawns a copy of itself through a hole in the sheath (this is the transcription). This copy folds back upon itself and builds its own sheath (this is the translation). After 151 time-steps, the automaton has acquired an offspring which looks exactly like its parent, and starts reproducing in turn. The parent has in the meantime been turned by 90 degrees, and starts spawning again. This time it assembles an offspring in the north, while the previous one had been built in the east. After four children, there is no room left for any more. The loop has blockaded itself, and can spawn no more. As a result, weirdly enough, its instruction gets erased: the automaton dies. But meanwhile, some of its offspring have become parents. Gradually, the whole plane gets covered by loops, most of them empty or, if we like, *dead*, but some others in different stages of their

```
2 2 2 2 2 2 2 2                              2 2 2 2 2 2 2 2
2 1 7 0 1 4 0 1 4 2                          2 4 0 1 1 1 1 7 2
2 0 2 2 2 2 2 2 0 2                          2 1 2 2 2 2 2 0 2
2 7 2         2 1 2                          2 0 2         2 1 2
2 1 2         2 1 2                          2 4 2         2 7 2
2 0 2         2 1 2                          2 1 2         2 0 2
2 7 2         2 1 2                          2 0 2         2 1 2
2 1 2 2 2 2 2 1 2 2 2 2                      2 7 2 2 2 2 2 7 2 2 2 2 2 2 2 2 2 2 2 2        2
2 0 7 1 0 7 1 0 7 1 1 1 1 1 2                2 1 0 7 1 0 7 1 0 7 1 0 7 1 0 7 1 1 1 1 2
2 2 2 2 2 2 2 2 2 2 2                        2 2 2 2 2 2 2 2 2 2 2 2 2 2 2 2 2 2

         Time = 0                                      Time = 35
```

```
2 2 2 2 2 2 2 2                              2 2 2 2 2 2 2 2
2 7 0 1 7 0 1 7 0 2           2 1 1 2        2 0 1 7 0 1 7 0 1 2        2 1 1 1 1 7 0 1 7 2
2 1 2 2 2 2 2 2 1 2           2 1 2          2 7 2 2 2 2 2 2 7 2        2 1 2 2 2 2 2 0 2
2 1 2         2 1 2           2 1 2          2 1 2         2 0 2        2           2 1 2
2 1 2         2 0 2 2 2 2     2 1 2          2 0 2         2 1 2 2 2 2 2            2 7 2
2 1 2         2 7 2           2 0 2          2 7 2         2 4 2                    2 0 2
2 1 2         2 7 2           2 0 2          2 7 2         2 0 2                    2 1 2
2 0 2 2 2 2 2 0 2 2 2 2 2 2 2 2 1 2          2 0 2 2 2 2 2 1 2 2 2 2 2 2 2 2 7 2
2 4 1 0 4 1 0 7 1 0 7 1 0 7 1 0 7 2          2 7 1 1 1 1 1 0 4 1 0 4 1 0 7 1 0 7 1 0 2
2 2 2 2 2 2 2 2 2 2 2 2 2 2 2                2 2 2 2 2 2 2 2 2 2 2 2 2 2 2 2 2

         Time = 70                                     Time = 105
```

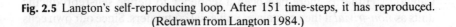

```
                                                     2
                                                   2 1 2
                                                   2 7 2
                                                   2 0 2
                                                   2 1 2
2 2 2 2 2 2 2 2         2 2 2 2 2 2 2 2      2 2 2 2 2 7 2          2 2 2 2 2 2 2 2
2 0 1 1 1 1 1 7 0 2     2 1 7 0 1 7 0 1 4 2  2 1 1 1 7 0 1 7 0 2   2 1 7 0 1 4 0 1 4 2
2 4 2 2 2 2 2 2 1 2     2 0 2 2 2 2 2 2 0 2  2 1 2 2 2 2 2 2 1 2   2 0 2 2 2 2 2 2 0 2
2 1 2         2 7 2 2 7 2         2 1 2      2 1 2         2 7 2   2 7 2         2 1 2
2 0 2         2 0 2           2 4 2          2 0 2         2 4 2 2 7 2         2 1 2
2 4 2         2 1 2 2 1 2     2 0 2          2 4 2         2 1 2           2 0 2         2 1 2
2 1 2         2 7 2           2 1 2          2 1 2         2 7 2 2 7 2         2 1 2
2 0 2 2 2 2 2 0 2 2 2 2 2 2 2 2 1 2          2 0 2 2 2 2 2 2 0 2  2 1 2 2 2 2 2 1 2 2 2 2 2
2 7 1 0 7 1 0 7 1 0 7 1 0 7 1 1 1 2          2 4 1 0 7 1 0 7 1 2   2 0 7 1 0 7 1 0 7 1 1 1 1 1 2
2 2 2 2 2 2 2 2 2 2 2 2 2 2 2                2 2 2 2 2 2 2        2 2 2 2 2 2 2 2 2 2 2

         Time = 120                                   Time = 151
```

Fig. 2.5 Langton's self-reproducing loop. After 151 time-steps, it has reproduced. (Redrawn from Langton 1984.)

life-cycle from budding looplets to spawning adults to decaying shells (Fig. 2.6).

Starting from this, John Byl devised self-reproducing automata which were still smaller.* The record is by now down to a configuration of 12 cells in a cellular automaton with six states and 57 transition rules. To an untutored eye, the fine distinction between transcribing and translating information seems barely visible any longer. Gone, too, seems to be John von Neumann's *complexity threshold* separating trivial from non-trivial self-replication. This is all to the good. A smooth transition fits in quite well with recent speculations on the origin of life, like the *genetic take-over* theory of Graham Cairns-Smith.* This theory postulates a series of gradual replacements from self-replicating clay minerals (crystal growth) to the ancestors of primitive cells. The clay crystals could use organic molecules, for instance sugars or nucleotides, for improving their own growth and division, and hence their propagation; this organic coat would then be subject to selection, and could

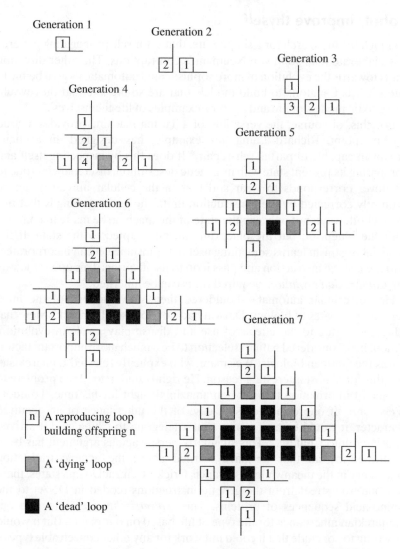

Fig. 2.6 The growth of a loop colony. (Redrawn from Langton 1984.)

evolve towards more and more efficiency, finally to become able to replicate by itself and get rid of the crystal altogether. In such a scenario, there would be a smooth transition from crystalline to cellular replication. Cairns-Smith compares it to a multi-fibred rope: 'none of the fibres has to stretch from one end to the other.'

Robot, improve thyself

So much for the search for a simpler life, that is for self-propagating patterns less elaborate than John von Neumann's contraptions. The other direction leads towards the evolution of more sophisticated automata, its goal being, to quote Chris Langton, 'to build models that are so life-like that they would cease to be models of life and become examples of life themselves'.*

For this, of course, the versatility of a Turing machine provides a good starting point. Richard Laing, for example, has designed an artificial organism capable of partial self-repair.* It does this by inspecting itself and comparing its present state with its genetic description; discrepancies (due to whatever corresponds to wear and tear in the cellular universe) can be promptly corrected by the construction unit. The exciting thing is that this works both ways: not only can the state of the machine be made to conform with the blueprint, but the blueprint can be adapted to the state. If the artificial organism learns something useful, for instance, it can incorporate it into the genetic instruction and pass it on to its offspring. This opens the way for *Lamarckian evolution*: acquired traits can be inherited.*

Hence, cellular automata shouldered their way into one of the most venerable debates of biology. Darwin had not opposed Lamarck's view that adaptations due to the effects of use and disuse play a role in evolution, although he considered natural selection to be a much more important factor. It was the German biologist Weismann who explicitly rejected Lamarckism, and this for a most interesting reason. He rightly understood the problem to be one of information transfer—an amazing insight for his time. To quote Weismann: 'If one came across a case of the inheritance of an acquired character, it would be as if a man sent a telegram to China, and it arrived translated into Chinese.'* This singularly perspicacious argument has been confirmed by all that we have learned about the genetic translation machinery in the meanwhile. As Francis Crick's Central Dogma states, there is a one-way street from the genetic instructions (coded in DNA) to the amino-acid sequences of proteins. This *irreversibility* seems to rule out Lamarckian inheritance for the type of life based on our planet. But it would be wrong to conclude that it could not work for any other conceivable type of life. Laing's automaton shows that it can, at least in principle.

Whether it would be of great value to inherit the effects of use and of disuse is another question. Speaking for myself, I would not care to pass on most of the traits which I have acquired during the last twenty years. I mention my shortsightedness, my taste for cigars and, why not, last week's haircut. (My wife could easily add a few more items to this list.) Most changes are not likely to be improvements. From this it has been argued that even if Lamarckian inheritance were possible, it could not lead to adaptive evolution. But Laing's automaton may be provided with some goal-seeking behaviour; it could evaluate the changes in its state and incorporate only those which look promising; it could even strive to evolve into higher and more complex forms. This inherent drive for self-improvement was also part of Lamarck's doctrine, and this is the part which Darwin rejected.* But we see that even such a strong version of Lamarckism can hold for artificial life. This is not to say that it will work any better than natural selection. In fact, it would be fun to devise evolutionary races between Darwinian and Lamarckian life-forms in a cellular automaton.

But this is science fiction. So is, for the moment, the possibility that we may alter the human genome in a purposeful way, which would be another realization of Lamarckism. Personally, I feel safer with the cellular automata. The great thing about them is that they can so easily be stopped. Escapism, maybe, but so what? Let us return, therefore, to the escape button, the early 1970s and Conway's game of *Life*.

Gunning for *Life*

The thing that got Conway started on the somewhat quixotic endeavour to look for life in the *Life*-universe was a surprising discovery by the young gang of enthusiasts from MIT who had devised the Glider Gun: namely, that a carefully orchestrated collision of 13 Gliders was able to produce this Gun.* On the display screen, it looks like a bit of magic: the crash of 13 Gliders gives birth to an infinity of them (Fig. 2.7). This was the spark that set a blaze going. Conway still had lots of tricky problems to solve, of course; but this, after all, was what he was here for.

John von Neumann's cellular automaton had 29 states; Codd's machine

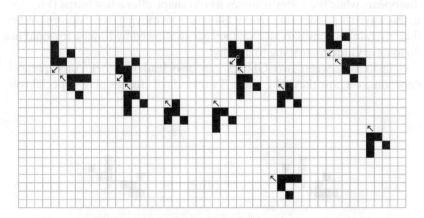

Fig. 2.7 Thirteen Gliders, four from north-east and nine from south-east, collide to build a Gun. The arrows denote the direction of movement.

had eight; Conway's *Life* game had only two states, the barest minimum. Both John von Neumann and Codd had provided their artificial worlds with laws intended to facilitate self-reproduction. In contrast to this, the simple laws of *Life* were not at all tailored to the task of permitting patterns to propagate. Among all possible worlds of cellular automata, *Life* was much more plausible than John von Neumann's or Codd's artefacts.

Conway's self-replicating pattern consists, once again, of a universal computer with a suitable program. The problem is how to build such a computer out of *on-* and *off-*pixels in a plane. But before we describe Conway's wonderful ways, we should eliminate some sources of semantic confusion. We are accustomed to think of a real computer behind the *Life*-screen, running the show. Such a computer has nothing whatsoever to do with Conway's two-dimensional version of the universal computer, which is going to be a pattern on the screen performing like a Turing machine ought to. The word *automaton* is also used in two distinct ways. The every-day type of automaton is a device for doing things, an object. The *self-reproducing* automaton Conway has in mind is in this sense an object of the *Life*-universe: a device for replicating itself. By an unfortunate coincidence, the *Life*-universe itself is also designed as an automaton: a *cellular* automaton, defined by its transition rules. We have to distinguish between the universe and the object in it; between the rules and the pattern.

This pattern will be assembled out of four types of building stones: Gliders, Guns, Blocks, and Eaters. The Gliders correspond to electrical impulses in computers of a more common type. They move along diagonals which correspond to the wires. Collisions of Gliders can produce not only Guns, but also Blocks and Eaters (Fig. 2.8). Eaters are small fishhook-shaped still-lifes. If a Glider bumps into an Eater in the right way, then the Glider disappears while the Eater resumes its old shape after a few burps (Fig. 2.9). In this way, it is possible to prevent Gliders from moving out of control. When their usefulness comes to an end, they get swallowed by Eaters. It would not do to have Gliders escaping to infinity.

Collisions between two Gliders can also lead to other consequences, depending on how they take place. Two kinds of crashes play a special role.

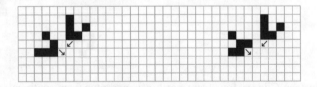

Fig. 2.8 Two Gliders build a Block (left); two Gliders build an Eater (right).

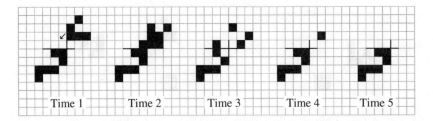

Fig. 2.9 An Eater devouring a Glider.

Two Gliders colliding at right angles can simply disappear (Fig. 2.10); this is called the *vanishing collision*. In the *kickback collision*, Gliders strike at right angles too: but only one of them vanishes, while the other Glider makes an about-turn and heads back to where it came from, its path shifted by half a diagonal square (Fig. 2.11).

And yet another class of collisions will be needed. A well-aimed Glider can make a Block disappear. Two properly spaced Gliders hitting a Block make it move closer by three diagonal squares (Fig. 2.12); and a fleet of 30 Gliders (yes, 30), all zooming in from the same direction, can push a Block back by

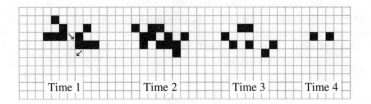

Fig. 2.10 A vanishing collision between two Gliders.

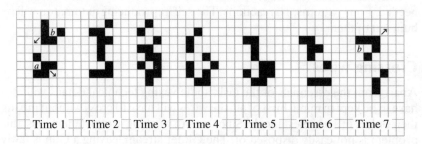

Fig. 2.11 A *kickback* collision between two Gliders: *a* vanishes while *b* reverses its course (its path shifted by half a diagonal square).

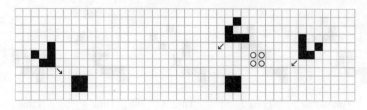

Fig. 2.12 A Glider blasts a Block (left). Two Gliders pull a Block closer (right). The circles denote the final position of the block. The Gliders vanish.

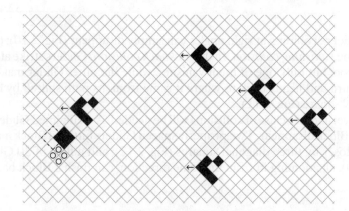

Fig. 2.13 Five Gliders push a Block to the position indicated by the circles. A further, symmetrically arranged fleet of five Gliders from the same direction pushes the Block to the dotted position. Altogether, the Block is shifted by one diagonal square. In this, and all the following figures, the screen is turned by 45 degrees.

three diagonal squares (Fig. 2.13). In these collisions with Blocks, the Gliders always vanish after their deed.

This, then, is the full list of elementary patterns and elementary reactions needed. Four patterns, 10 reactions. The self-reproducing automaton will be built out of them.

Crossing wires

Apart from the essentials of the task, some vexing technical details will also have to be kept in mind. First of all, the wires will have to cross without mixing their pulses in a hopeless tangle. Since we are working in two dimensions, this problem is not easily disposed of. The Glider stream leaving a Gun is fairly dense: every 30 time-steps a new Glider passes by. If it crosses another stream, collisions will take place. But with three Guns and one Eater, a so-called *Thin Gun* can be fixed up, whose Glider stream is thinned out to any desired degree (Fig. 2.14). For convenience, we shall turn the screen by 45 degrees, so that the wires run north–south and east–west.

From now on, our Guns will all be thin. Their firing sequence beats the measure for the machine. It is the internal clock of the two-dimensional computer; one that, incidentally, will run to a very slow time.

For transmitting information coded by sequences of zeros and ones, we have only to delete or preserve the corresponding Gliders in the thinned-out streams. Messages which are properly timed can cross now without interfering with each other.

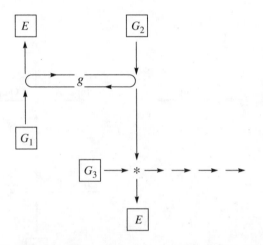

Fig. 2.14 The Thin Gun. The Glider *g* moves back and forth between the tracks of the Guns G_1 and G_2, and destroys (for example) every four-hundredth Glider from the streams leaving G_1 and G_2. Hence only every four-hundredth Glider from G_3 can escape. (*E* is an Eater, ∗ is a vanishing collision between two Gliders.)

Next to the question of how to cross wires, there are three further problems which have to be solved if we want to mimic the circuitry of a real computer. (a) How can we bend a wire by 90 degrees? (b) How can we split a wire in two, both carrying the same message? And (c) how can we collect several parallel wires in a tight bundle? For the usual computers made of silicon chips and electrical wires, these are not problems at all, but for its *Life*-like imitations they are fairly hard. How, for example, are we to send off closely spaced squadrons of Gliders without having Guns hitting other Guns? We shall have to return to these questions later.

Compared with this, the construction of the *logical parts* of the computer seems simplicity itself. We need only two logical gates, corresponding to the elementary operations *and* and *not*. Every logical operation used by the computer can be generated by means of negations and conjunctions. The disjunction *A or B*, for example, is simply the negation of *not-A and not-B*.

The *not*-gate transforms a message into its complement: a sequence like 10110... turns into 01001... This can be obtained by sending a continuous stream of Gliders to intercept the input message at right angles, timing it so that the collisions annihilate both Gliders (Fig. 2.15). In this way, every approaching Glider gets destroyed, while only those Gliders from the intercepting Gun which do not hit any incoming target manage to escape. Hence there will be a Glider in the output if and only if there was no Glider in the input.

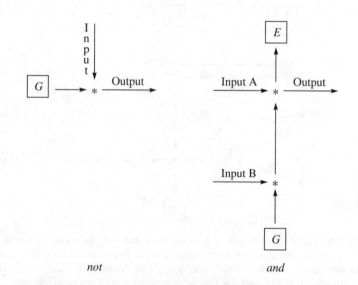

not   and

Fig. 2.15 The logical gates *not* and *and*. The Guns fire continuously. All collisions between two Gliders are of the vanishing type. (*G* denotes a Gun and *E* an Eater.)

The *and*-gate transforms two input messages tracked in parallel into one output. It lets a Glider go whenever both incoming streams carry Gliders. The two messages 0110101... and 0011010... yield in conjunction 0010000... This kind of gate is obtained by one Gun firing at right angles through the two input streams.

You will have noted that the output from a *not*-gate moves at right angles to the input. Using four negations which cancel pairwise, we can delay the original message by as much as we like, simply through routing it into a blind alley and back. In this way we can hold a message until we need it. By now, we can also see how Conway solved two of the technical problems previously mentioned: how to bend a wire, and how to split it. This is shown in Fig. 2.16.

Apart from the internal memory of the universal computer, which we may picture as a very, very long message running round and round in a closed loop, we also need an external memory capable of extending, in principle, to infinity. In the usual model of a universal computer, this corresponds to the infinite tape. Alternatively, one might use some form of register able to store arbitrarily large numbers. Every piece of information can be coded into some number. (The blueprint for the self-reproducing automaton will be a number of staggering size, of course.)

Conway decided to represent such a number by the *distance* of some Block to the universal computer. The computer will therefore need to be able to do three things: (a) check whether the distance is zero; (b) reduce the distance by one unit; (c) increase the distance by one unit. Testing for zero is simple: one just has to shoot a Glider transversely at the position with register value zero. If a Block happens to sit there, then Block and Glider both vanish. If not, then the Glider emerges unharmed and will be assimilated by the internal memory. Next the computer sends a squadron of two Gliders out towards the Block, and pulls it closer by one unit (three diagonal squares); tests for zero again, and so on . . . In order not to lose the stored information by calling it up, the Block of another register of the external memory has to be pushed away by the same amount, a task which, as we know, requires per unit a carefully arranged flotilla of 30 Gliders. Incidentally, the time intervals between consecutive operations in a register have to get larger and larger, or else the operations would mix up. This is a terribly ponderous way of reading the memory: it takes literally ages. But time is in ample supply: *Life*-computers are forever.

Since all the *hardware* of this universal computer (made of Guns, Blocks, and Eaters) can be obtained from colliding Gliders, it can be manufactured by rigging up a well-planned crash between four huge Glider fleets converging from the four points of the compass. Of course the different collisions need not happen simultaneously. The fleets will be widely scattered over the plane. The Eaters will be built before the Guns, to liquidate all superfluous Gliders. Again, there is no need for haste; and no want of space either.

Alive and kicking

Now for the last difficulty. The self-reproducing automaton will not be able to nurture offspring in its interior, like a mammal. It has to arrange the collisions somewhere out there in the open spaces. But how, then, can it guide the four armadas upon converging courses? The Gliders leaving the computer will sail further and further away. They cannot be thrown like boomerangs, and reverse direction midway. Or can they?

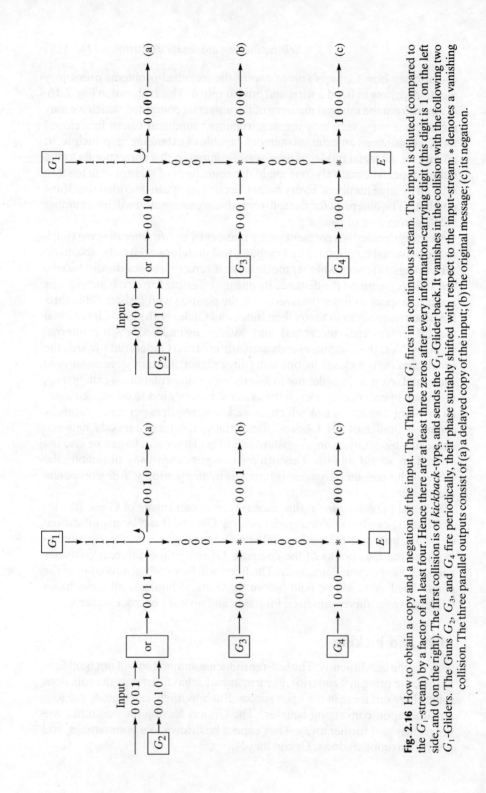

Fig. 2.16 How to obtain a copy and a negation of the input. The Thin Thin Gun G_1 fires in a continuous stream. The input is diluted (compared to the G_1-stream) by a factor of at least four. Hence there are at least three zeros after every information-carrying digit (this digit is 1 on the left side, and 0 on the right). The first collision is of *kickback*-type, and sends the G_1-Glider back. It vanishes in the collision with the following two G_1-Gliders. The Guns G_2, G_3, and G_4 fire periodically, their phase suitably shifted with respect to the input-stream. * denotes a vanishing collision. The three parallel outputs consist of (a) a delayed copy of the input; (b) the original message; (c) its negation.

Well, they can. Conway found a neat trick (Fig. 2.17). It is based on kick-back collisions, which, as we know, get rid of one Glider and send the other Glider back on a closely parallel track. It uses three streams of messages (and hence three pre-programmed Guns). We may assume that the Gun whose information we wish to transmit fires from west to east. The other two Gliders should be pointing north, traversing its path. A Glider belonging to the message sent eastward will be allowed to cross the line of the first transverse Gun without hindrance; but it will be sent back by the second Gun, and then out again by the first, back by the second, and so on, and so on in a long series of kickback collisions. At each turn, its track gets shifted slightly to the north.

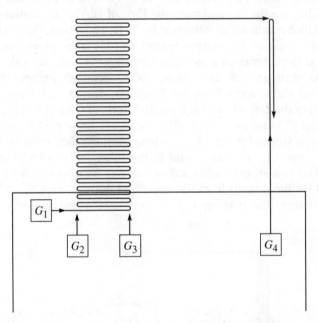

Fig. 2.17 *Sidetracking.* G_1 fires the input sequence, G_2 and G_3 shift it by means of *kickback*-collisions (which destroy the Gliders from G_2 and G_3). On the right, *double sidetracking* is in action: the message fired off by G_4 is sent back by *kickback*. The guns have to be programmed to fire in the right sequence.

As on a mountain road with many hairpin bends, the Glider winds its way upwards, slowly but steadily moving away from the computer. At some suitable time the second transverse Gun stops firing and lets the Glider escape. The Glider now moves to the east, its original direction, but on a track which has been shifted by an arbitrary amount to the north. Conway calls this *sidetracking.* And if this Glider meets a Glider from yet another Gun pointing north in a well-timed kickback collision, then it sends the upstart back to the south, and hence towards the computer, while discreetly dissolving itself. This *double sidetracking* does it: it yields the desired boomerang.

It is this *sidetracking*, by the way, which solves the last technical problem: how to send Gliders off along tightly bundled wires without knocking down one's own Guns. The Gliders, as we now see, can all originate from one and the same Gun, and be distributed by some transverse barrage along several closely packed tracks.

So now we can watch an automaton giving birth (Fig. 2.18). It will be contained in some huge rectangle, its corners bristling with many triplets of side-tracking Guns. Somewhere far out to the south, there will be a few Blocks. They are the external memory. First of all, the automaton reads the register which contains the instruction for its reproduction. Then its universal computer sets up all necessary preparations (this takes a very long rumination). Next the automaton grows arms; outriggers reach out in slow motion far to the north and the east. Along them, seemingly endless streams of Gliders will zigzag away from the computer, deeper and deeper into the emptiness of the *Life*-plane. Then the first fleets of Gliders will escape from the outriggers: those from the northern arm will move east, those from the eastern arm will sail north. After a long journey, they meet in *kickback* collisions; half of the Gliders vanish, the other half doubles back on its course. The two armadas obtained in this way by *double sidetracking* are intended for the inroads from the north and the east. Two other fleets are taking off from the outriggers now; the western and southern armadas

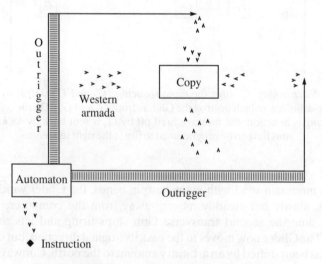

Fig. 2.18 The automaton reproduces itself.

(obtained by simple *sidetracking*) are on their way. The outriggers have done their job, and dissolve into thin air. Meanwhile, in a huge rectangle to the north-east of the mother automaton, the collision of four worlds produces a new automaton. Only one thing remains to be done: the instruction must be encoded into a number, and handed over to Junior's external memory by carefully crashing Gliders to have Blocks stand at their proper place. And here we are—over and out.

Ars longa, vita brevis

Such an automaton, by the way, can not only propagate itself. It can also kill itself. Properly aimed Gliders can annihilate Blocks, Guns, and even Eaters, destroying themselves by the same stroke. The automaton has only to send off four well-assembled fleets and have them reverse their course through *double sidetracking*. The fleets return to sender, converging upon the automaton and wiping it off (Guns first, Eaters last) until no trace is left. This seems a pretty drastic measure, but it offers a way for the automaton to move: it can build a copy of itself on a vacant lot of its choice, and then gracefully dissolve.

Of course it can do a good many other things, too. After all, it is the proud owner of a universal computer. In particular, it can tackle any arithmetical problem, like searching for the solutions of $x^n + y^n = z^n$ in whole numbers x, y, z, and n. According to Fermat, there are no such solutions for n larger than 2; but we do have to take his word for it. Mathematicians have laboured for more than three hundred years without finding a proof of this conjecture.* We could ask the *Life*-computer to check methodically for solutions. Conway proposes that if it succeeds, it should tell us so by ejecting x Gliders to the north, y to the east, z to the south, and n to the west, and then commit hara-kiri. I fail to see the use of that suicide—after all, a counter-example to Fermat's conjecture needs no publicity stunt. But possibly the computer wants to avoid having next to check Goldbach's conjecture that every even number is the sum of two primes.

Anyway, the point is this: we do not know, today, whether the automaton will succeed in its quest for a solution to Fermat's equation or not; that is, whether its program will ever halt. We may possibly know it tomorrow. But as Turing has shown in his solution of the so-called decision problem, there is *no general procedure* permitting us to predict, for any given program, whether or not it will ever halt.* *Life*, in this sense, is unpredictable, in spite of its simple transition rules: unpredictable not just because of the deficient wetware which we carry in our skulls, but fundamentally.

Life, the universe, and everything

One should be warned that whoever wishes to witness a suicide or a happy event in his or her *Life*-time is liable to need the utmost patience. Conway's mills work very slowly. His automaton consists mostly of gaping emptiness. Only rarely will one encounter a Glider passing through the night, and hardly ever find any fixed part (Block, Eater, or Gun). The time and space needed for its replication defy imagination. But *our* space and *our* time (and also *our* imagination, of course) do not belong to the *Life*-universe. They are external references. *Life* ticks at its own rate. Besides, our own life is also very slow, compared with the speed of biochemical reactions in our bodies.

Now let us stretch our imagination somewhat more, and ask about the origin of life in the *Life*-universe. Conway's answer is that it is inescapable. Inescapable, at least, under certain rather mild assumptions. If the game board is totally empty, of course, then nothing will ever happen; if, conversely, every cell is occupied, then all will be polished off in the next instant; and similarly, there will be countless other initial positions yielding no self-replicating patterns. But these are all special cases. We should not try to tamper too much with *Life*. The simplest way to distribute matter *randomly* in the plane would be to have every cell, independently of its neighbours, occupied or not with some given probability. In this case, every conceivable finite pattern will certainly occur *somewhere* in the infinite plane. (It will even occur infinitely often.) Self-reproducing patterns will therefore also be bound to occur; very sparsely, to be sure, but present nevertheless. In this sense, the world of *Life* contains life almost as soon as it contains matter.

Many very different configurations will have the property of self-reproduction. They will be submitted to a natural selection of sorts. Some will multiply faster than others. Some will be quickly destroyed by Gliders happening to pass by, or by other patterns crawling across the plane. Some will be more resilient, or simply more lucky. Some will end up suffocated by their own offspring. Some will move too slowly, and some too fast. The proportion of successful patterns is bound to increase. The lifeless environment will change, too.

Some automata will evolve rudimentary sense organs to obtain information about their surroundings. Some will develop the faculty to move into the direction which appears the most promising, or to flee from dangers. Some will withdraw into shell-like structures, and some will evolve offensive weapons. There will be species exploiting others, and species set upon cooperation. In due time, multicellular beings are apt to emerge—huge colonies of automata, obeying a common program and begetting other colonies. In order to discover better blueprints, such automata may start to recombine their instructions, using some two-dimensional forms of sexuality.

There will be complex types of social interaction, and sooner or later some kind of intelligence too. These patterns will learn to feel and to think.

Once conscious beings have evolved, how will they feel about it all? How would they view their world? A nice little exercise for philosophers. What geometry would these Flatlanders be taught at school? What science? Could they ever grasp what *motion* means in their plane? Would they develop automata? Could they ever find out the basic laws of their world, Conway's neighbourhood rules? And will they be crazy enough to wonder if a computer was hidden behind their world, part of an altogether different universe? Will they speculate on whether the Supreme Programmer had created life, or only matter? And on whether He had had any freedom in choosing the laws of the universe? Will they sometimes feel doubts about whether they mean more to Him than a few specks of dust on the *Life*-plane? And will they fear that He might suddenly push the escape button to switch off the game—possibly because the Supervisor enters?

3

Population ecology and chaos

3

Population ecology and chaos
Playing cat and mouse, and other pursuits

> *Truth is the cry of all and the game of a few.*
> Bishop Berkeley

Screen flicks

We speak of the *moves* of a game, and of being *moved* by a play, even if we keep sitting throughout. These expressions show how intimately games are associated with motion. In particular, games of pursuit and evasion must have been played long before human children started running around. Mice, I think, can confirm this.

It takes a strong stomach to watch a cat at play with its victim. But on a computer screen, games of pursuit and evasion become clean fun. Take, for instance, that game of Shark and Fish introduced by A. K. Dewdney in his column on Computer Recreations in the *Scientific American*.* Like *Life*, it is a spectator game. Essentially, one has only to set it in motion and watch.

A certain number of fish move aimlessly over the screen, each one performing a random walk, but prevented by the rules from bumping into another; so that a fish with neighbours to the north, east, south, and west has to wait till the traffic jam clears up. From time to time, some fish breed by simply leaving newborn fish behind. So the fish population grows and would end up jamming the screen, if it were not for the sharks.

Sharks also mill around randomly and breed at regular intervals. They too are prevented from bumping into each other. But they are not forbidden to bump into the fish. On the contrary, that's how they feed on them. If they don't find enough to eat, they die after a while and vanish without a flicker from the screen.

So we can follow the fate of the fish and their predators. Dewdney describes how they wage their ecological war on the planet Wa-Tor. This planet has the shape of a torus (or doughnut): a fish moving east reappears on the west of the screen, a fish moving north in the south. This shape, unusual as planets go, is most convenient for a screen display.

Sometimes the ocean of Wa-Tor becomes so crowded with fish that the sharks multiply until they wipe the plate clean. They will, of course, all die after such an excessive feast. On other occasions, the sharks aggregate in one region and eventually starve without noticing that a small school of fish has subsisted elsewhere. In this case, the surviving fish, being rid of the sharks, can cover the whole ocean until it is packed as solid as a tin of sardines. But if the parameters are well chosen (the breeding time of a shark three times that of a fish, for instance), then the fish and the sharks can coexist for ages. Their populations are apt to fluctuate wildly up and down, but they are fairly safe from extinction: whenever the sharks are few, the fish multiply to the point where some sharks are bound to stumble upon them and to thrive; conversely, whenever the fish are few, the sharks become rare and stop pestering them. The population numbers tend to cycle in time, the prey curve dodging up and down, with the predator curve in hot pursuit (see Fig. 3.1). On the screen, distributions condense and disperse, each school of fish being assailed by more and more sharks until the balance tilts and the local patch crashes.

Needless to say, the feeding and breeding behaviours could be greatly refined by any talented programmer. The fish could be equipped with strategies for dispersal and evasion, and the sharks with methods for searching more efficiently. One could even have these programs evolve: from time to time, a random mutation could introduce some new strategy to be tried by 'natural' selection. We could even hope to see the fish split up into different species, one opting for tight schools, say, and the other for roaming freely around; and the sharks might specialize too, some searching in narrow circles and others in wide sweeps. Some might even switch their foraging habits according to circumstance. For the moment, however, this is computer

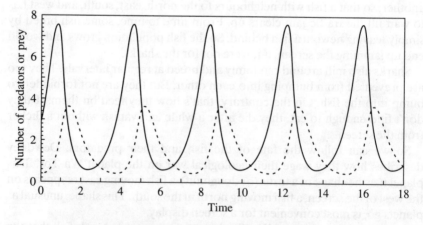

Fig. 3.1 A computer solution of the Lotka–Volterra equation. The numbers of predators (dashed line) and prey (solid line) oscillate.

fiction. On the planet Wa-Tor, such fine-tuning of genetic instructions is not part of the game. But what about planet Earth?

War games

War, some ancient said, is the father of all things. An exaggeration, without doubt. But it must be admitted that war made a big contribution to the birth of population ecology.

During the First World War, the Adriatic was what military writers call a secondary theatre, but the struggle between the navies of Italy and the Austro-Hungarian Empire effectively put a stop to any large-scale fishing in the region. This had an unexpected side-effect, which was noticed by the Italian biologist D'Ancona when he investigated the statistics of fish markets, some years later: the proportion of predatory fishes appeared to be considerably higher, during the war, than in the years before and after. D'Ancona was puzzled. Why should war favour sharks? It couldn't be just the *Zeitgeist*, surely.

D'Ancona put the question to his prospective father-in-law, Senator Vito Volterra,* who held the Chair of Mathematical Physics in Rome and was widely recognized as one of the leading mathematicians of his time. See here, said Volterra, if we denote by x and y the numbers of prey and predators . . . And he proceeded to a differential equation which, amazingly enough, gave a convincing answer to D'Ancona's problem.

Some fish will have a large number of offspring, and some none at all. If we take the average, we get the *growth rate* of the population: the mean individual contribution to population growth. There are many species of fish in the Adriatic, even today, but for simplicity, we suppose that there are only two, predator and prey. This gives us two rates of growth; and Volterra observed that each depended on the size of the *other* party. Every supplementary predator reduces the growth rate of the prey: so, if the number of predators exceeds a given threshold value, the prey's population size will drop. Conversely, predators will do better if their prey gets more numerous. If the number of prey is below a given threshold, then a predator is more likely to die of hunger than to reproduce, and its population will wane.

If both population numbers happen to be *exactly* as large as the corresponding threshold values, neither will have to change. The system, then, is at rest, or *in equilibrium*. The numbers balance out. But any small perturbation will set the balance swinging. With a few prey fish less, for instance, the size of the predator population will drop; which is all to the good for their prey, whose number will grow. The predators will then cash in and their numbers will grow in turn. As soon as it exceeds the threshold value, the amount of available prey drops, and so on. Hence, oscillations. The population numbers will sometimes be above the threshold values and sometimes below. But on average, this cancels out; the averages are just the threshold values, in fact.

Now for the fishermen to join the game. In their nets, both predators and prey can get caught. This reduces their rate of growth, but it affects their threshold values in opposite ways. For the prey, it becomes harder to reproduce. A *smaller* amount of predators will suffice, now, to keep them in check. Hence, the mean population size of predators will drop. For the predators, it also becomes harder to reproduce. They need a *larger* amount of prey than before. Thus the mean population size of prey will increase, rather surprisingly. If fishing is stopped, as it was during the war, the contrary happens: more predators, fewer prey. Which is just what D'Ancona had observed and what had to be proved.

This became known as *Volterra's principle*: whenever the population sizes of prey and predator determine each other, a decrease in their growth rate (caused by hunting, for instance) leads to an increase of the prey and a decrease of the predator. The principle has been confirmed by some catastrophic failures in insect pest control.* The pesticides usually act not just on the so-called pest (caterpillars, for instance, or aphids), but on their natural enemies too (beetles or birds). The action on these predators is even enhanced, since they are higher up in the food chain and get their poison in concentrated form. The pest's generation time, moreover, tends to be shorter, so that it can adapt more rapidly. But even without this genetic backlash, an insecticide campaign frequently leads to an increase of the pest, simply by the depletion of their natural opposition.

It turned out that Volterra's model had already been studied—albeit not at the same depth—by the American Alfred J. Lotka, a maverick scientist of astonishing versatility. Lotka had been inspired by problems in epidemiology and chemistry, but again, war played a role in shaping his thoughts. He found it 'well worth considering whether interesting light may not be thrown on various problems of biological conflicts ... somewhat after the manner in which the war game imitates the armed conflict of nations'.*

Few ecologists were happy with Volterra's model (apart from D'Ancona, of course). They deemed it too simple, and so it was indeed. Volterra refined it a great deal, and devoted the last years of his rich career to mathematical ecology. But biologists kept complaining that his writings were (I quote) 'dry as dust and almost example-free'.* Mathematicians had fewer objections. Three brilliant Russians, Gause, Kostizin, and Kolmogoroff, extended Volterra's work,* and showed in great generality that there were only three (well, maybe four) possible outcomes for the ecological warfare between predators and prey.

It may—this is the first case—so happen that not even the smallest group of predators is able to make a living out of what prey their habitat can sustain. This occurs when the predator is not efficient enough in turning captured

prey into offspring. If its efficiency were higher, it could reach a stable sub-
sistence level. In this second case, prey and predator numbers tend to reach
well-specified equilibrium values. If these values were perturbed, the population sizes would reassume them after some damped oscillations, exactly as a
pendulum would return to rest. But if the predator is still more efficient, the
two population sizes will keep oscillating and never settle to rest. This is the
third case: the numbers of predator and prey cycle as regularly as an
alternating electrical current.*

All this is theory, of course: in reality, things will be messed up by random
events. These are particularly dangerous if the population is small: a few
accidents then can suffice to wipe it out. This is the fourth outcome alluded to.
If a population cycles up and down, it will periodically be on the low side and
court disaster. In this sense, systems with efficient predators are more vulnerable to chance. Intuitively, this is not surprising: a stable coexistence is not
possible if each eaten prey turns into many greedy young predators on the look-
out for more. What seems strange, however, is the paradox of enrichment:* if
an environmental change makes the habitat support *more* prey, this can turn
an equilibrium into a cycle, and hence *reduce* the stability of the system. This
may happen, for instance, if a surplus of fertilizer is drained into a lake.

We have met with fluctuating numbers on the planet Wa-Tor. They can be
found in predators and prey down here on Earth, too. Our most impressive
records are provided by the venerable Hudson's Bay Company. For more
than two centuries, it has kept books on the catches of Canadian trappers. It
turns out that the numbers of lynxes and snowshoe hares (the latter a
favourite dish of the former) peak more or less regularly at 10-year intervals.
Many textbooks on ecology display their statistics (see Fig. 3.2) as a striking
proof of the validity of the mathematical models. But some ecologists
quibble. They frown. They raise doubts.*

Do lynxes let their hare down?

One thing sceptics are quick to point out is that the trappers were not intend-
ing to take a census of Canadian wildlife. They needed the furs to make a
living: their catches might well reflect economic rather than ecological cycles.
It is well known that the market mechanisms of supply and demand tend to
oscillate, particularly in a field so closely tied to fashion. So the pelts could
simply reflect the changing objectives of the trappers. But even if they faithfully mirrored oscillations in the population densities, these need not
necessarily be caused by predation: periodic outbursts of disease, for
instance, could also account for it.

Are lynxes able to affect appreciably the number of hares? The answer is
not obvious. Indeed, it appears that those mammals which come first to a
layperson's mind when predators are mentioned (lions or wolves, for

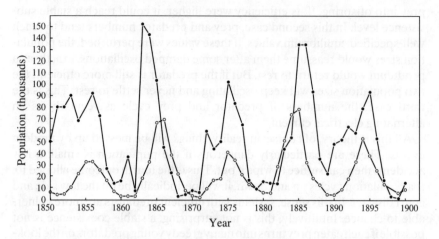

Fig. 3.2 The number of lynxes and hares trapped for the Hudson's Bay Company. The black dots denote the hares; the white dots, the lynxes. This is surely the best-known graph in ecology.

instance) are not so staggeringly superior to their prey as to be able to deplete them in a big way. A wildebeest is not easily brought down by a lion, and musk oxen or caribou have frequently been observed to stave off attacks by packs of hungry wolves. These predators seem reduced to picking their prey from among the elderly and the sick, and hence to kill only those doomed to die soon anyway. The British ecologist Colinvaux writes that such predators ought to be seen as 'scavengers without the patience to wait for their meat to die'.* A lion must feel calumnied, I guess. But it is certainly true that among spiders and insects, one finds incomparably more efficient hunters and killers—so efficient, in fact, that their very success can lead them to the verge of extinction, or beyond.

The status of lynxes was for a long time under dispute. But now it seems that they are rather good at their job, after all. The best evidence so far comes from a mathematical model, and a rough estimate of the relevant quantities entering it—the maximum number of prey which can be sustained by the habitat, and the growth rates under optimal conditions. The question, then, is whether the model will cycle for these values. In a review of eight pairs of predators and prey (like muskrat–mink or moose–wolf), the lynx–hare team emerged as the only couple likely to keep cycling.* This is no direct proof— the lynx is not caught red-handed, so to speak, in the act of decimating hares—but direct proof is frustratingly difficult to obtain.

How difficult it is can be seen from field studies in Newfoundland, which managed to establish that lynxes do regulate the size of caribou populations.*

(Not quite the same thing as regulating hares, but a step in the right direction.) The lynxes' effect upon adult caribou is nearly negligible, but their impact upon caribou calves turned out to be huge; and this, of course, reduces population growth much more effectively than killing off the old. For years, the main cause of death for caribou calves had been known to be septicaemia caused by abscesses in the neck. Then, a calf was found with four puncture wounds in the neck, and the full truth came out: these deadly abscesses were due to wounds inflicted by the lynxes' fangs. This was proven beyond reasonable doubt when the bacteria causing septicaemia were found in the saliva of lynxes.

It must be vexing for a biologist to find that mathematicians who were extolling the role of the lynx in making hare numbers oscillate will tend, on being told that lynxes do *not* deplete hares, not to challenge this claim, but rather to come up with a different model producing the same oscillations of hares without any recourse to predators. Compared with the tasks of ferreting scattered remains of caribou calves out of the undergrowth, or of sampling the spit of a lynx, it seems like a sleight of hand. But the maddening thing is that the theoreticians' fall-back solution has the same kind of plausibility as the prey–predator mechanism. It could be that the hare population cycles all by itself, and that the lynxes, who to a large extent depend on the hare for their fare, simply follow its ups and downs with an appropriate delay, but without by any means *causing* them.

So what *is* causing them, in that case? Simply self-regulation, maybe. A given habitat offers accommodation for only so many hares; this population size corresponds to a balance between birth and death rates. If the population is smaller than this value, it will grow; if it is larger, then it will diminish. This suggests that it has to converge towards its equilibrium value—the habitat's full capacity. But one need not be a control engineer to know that a feedback device of this sort can lead to oscillations instead. It is like regulating the water temperature under a shower; one is apt to overshoot, and get scalded, then to overreact and get chilled, etc.

The culprit is time-delay. If the response time is slow, it becomes difficult to adjust the taps properly. Such time-delays act upon populations too, in varying degrees. Many short-lived insects have separate generations, for instance: the parents never get to see their offspring. If this year's population is below full capacity, then each insect enjoys a surplus of resources and can lay lots of eggs. The insects hatching next year will be born into an overcrowded world, will not be able to find sufficient food, will rarely live to maturity, and will therefore leave only a few descendants. Similar fluctuations have been recorded in a great many free-living populations.* The population peaks of lemmings, which recur every fourth year, became proverbial. (The legend that they then collectively jump from a convenient cliff has been fostered by Hollywood filmmakers.)

In order to exclude interactions with other populations, ecologists have followed the fortunes of laboratory populations (for instance, of flies kept in jars*). These experiments, which go back to A. J. Nicholson in the 1930s, show violent oscillations without any tendency to dampen down: within 50 days, fly densities boom and bust, from a few hundreds to almost ten thousand and back (Fig. 3.3). So here it is, all bottled up, so to speak: the population whose numbers keep cycling, driven not by predation but by intrinsic dynamics. It almost reminds one of a cat which, in the absence of mice to play with, will chase its own tail round and round.

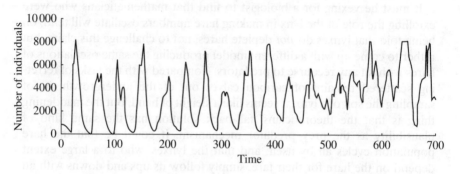

Fig. 3.3 The population numbers for one of Nicholson's fly populations (time in days).

Humps and jumps

In modelling population dynamics, there are two complementary lines of attack. One approach keeps very close to the species under investigation, renders a detailed account of its life cycle, reckons with the influences of temperature, disease, and age structure, and faithfully mimics the interplay with the other populations of the ecosystem. This yields a very complicated model and—usually—a very complicated behaviour. By comparing with the actual data on population densities, one can check if one is right on course or if some essential ingredient has been overlooked. The latter frequently happens, of course. But even if the predictions agree with the actual outcome, such a model will only provide for computer simulations which are no more enlightening than biological reality itself.

The other approach consists of modelling only the barest essentials of the system. One then has to dispense with predicting the actual growth of a given population; but one can try instead to gain insight into the basic features of its dynamics. The art, as so often, consists of omitting. A model does not get any better if it is loaded with detail. A sketch meant to show us a way out of the woods should skip lightly over the trees. Here is a Wittgenstein quote* to support this: 'Is there always an advantage in replacing a blurred image with a sharp-focused picture? Isn't the blurring frequently just what one needs?' And here an echo from Robert MacArthur, who dominated ecology in the 1950s and 1960s: 'These kinds of general events are only seen by ecologists with rather blurred vision.'*

So let's blur. Let's forget about a thousand details concerning the age pyramid, the climate, or the patchiness of resources. Let's assume that the population size y is entirely determined by the population size x from the previous generation. If we know how y depends on x, we know the whole future evolution of the model. We can program a computer to display y whenever we enter x. It is like playing tennis against a wall. We serve x and the computer returns y. Its return is a function of x: one usually writes $y = f(x)$. We can study this return by testing it for different values of x, and plotting the answers. After some warming-up exercises, we may decide to feed it on its own return. If we send the y-answer back into the computer, it returns $f(y)$, or $f(f(x))$. This is what a population of size x becomes after *two* generations (two bumps against the wall). If we enter this number in turn, we obtain the size after three generations, and so on. We may program the computer to feed itself upon its own output, and have it churn away the sequence of population sizes for as many generations as we like.

The outcome depends on the function $y = f(x)$ which describes how one generation determines the next. What do we know about it? One thing for sure: that it is *non*-linear. If doubling x always meant doubling y, there would soon be, as Darwin wrote, 'literally no standing room for the progeny.'* We obviously have to assume that the rate of growth decreases with population size. As long as the total is below some critical value, every supplementary individual will cause an increase in the next generation, but the returns will diminish. Once the total is above the critical value, every surplus member today will make for a smaller number tomorrow. This means that the function $y = f(x)$ is described not by a straight line, but by a curve with a hump, whose highest point is attained for the critical value. As long as x is below this critical value, y increases with x; and if x is above, then y decreases with x.

There are lots of single-humped functions. The simplest is the parabola, the curve describing the path of a stone thrown into the air. So let's try the parabola. The steepness of its hump depends on the angle at which the stone is thrown. This number corresponds to the highest possible growth rate, which is that of the smallest population. It is given by the fertility of a female living under optimal conditions, unfettered by any competition (very small x). Now if this fertility were too small, the population would die out. No need so far for a computer. But it comes in handy to show that for higher fertilities, the population approaches an equilibrium number corresponding to the capacity of the environment; and that for higher fertilities still, it displays oscillations which eventually dampen down, so that the final outcome is again a steady equilibrium value. If the fertility is somewhat higher still, we see that the oscillations persist: after an initial phase, the population numbers keep jumping up and down, alternating endlessly between the same two numbers. In all these cases, the size of the population we started with plays no role in the end: after a few generations, all initial differences are forgotten (Fig. 3.4).

So the model has done its duty, and done it well, in spite of its extreme simplicity. It shows that a population regulating its own size through competition can oscillate both in a damped or a protracted fashion. Increased fertility (caused, for instance, by higher temperatures) tends to destabilize the steady state and to replace it by stable cycling. This has actually been observed for water fleas kept in a jar: * at 18°C, their numbers settle at an equilibrium value, while at 25°C they keep jumping up and down (the numbers, I mean, not the fleas). One may record with pleasure that common sense has been confirmed, and turn with bolstered confidence to the modelling of more realistic cases. For it should be obvious that one cannot expect any more insights from the parabolic model of population dynamics, reduced as it is to a caricature.

One Australian physicist, however, who (like the chemist Lotka and the mathematician Volterra) had been captivated by ecology, was not ready to dismiss the parabolic population model in such a cavalier fashion. Robert May kept playing with it on his computer. What he found became a major impulse for a paradigm shift that, over the last few decades, has influenced all branches of science in a spectacular way. May found a road to chaos.

The demon and the butterfly

After a change in scale, the model equation sends x into $y = rx(1-x)$. The input x and the output y are now numbers between 0 and 1. The constant r corresponds to the maximal fertility, the unchecked rate of growth of a population too small to hamper its own expansion. With x_n, we denote the population size in the nth generation. Thus x_0 is the initial population size. In the next generation, it is x_1, which is nothing but $rx_0(1-x_0)$. Next comes x_2, which is $rx_1(1-x_1)$; if we plug in the value for x_1, we can write it as a function of x_0. In this way, the size of *any* generation is completely specified by x_0, but the formula expressing x_{15} as a function of x_0 would already fill this whole book. It is better instead to build a closed loop into the computer program: it can repeat the calculation steps leading from one generation to the next (one subtraction, two multiplications) as often as we like, so that we have only to lean back and watch the x_n run their course (Fig. 3.4).

All depends, as we know, on the value of r. If it is smaller than 1, the population size decreases steadily to 0. If it is between 1 and 3 (included), it approaches, as generations follow one another, a well-defined

	(a) $r = 2.3$		(b) $r = 3.2$			(c) $r = 4.0$
$x_0 = 0.23456000$	$x_0 = 0.23456000$	$x_0 = 0.78234000$		$x_0 = 0.98765000$	$x_0 = 0.98765001$
$x_1 = 0.41294568$	$x_1 = 0.57453314$	$x_1 = 0.54490922$		$x_1 = 0.04878991$	$x_1 = 0.04878987$
$x_2 = 0.55756954$	$x_2 = 0.78222340$	$x_2 = 0.79354612$		$x_2 = 0.18563782$	$x_2 = 0.18563768$
$x_3 = 0.56737721$	$x_3 = 0.54511986$	$x_3 = 0.52425816$		$x_3 = 0.60470568$	$x_3 = 0.60470532$
$x_4 = 0.56455871$	$x_4 = 0.79348544$	$x_4 = 0.79811693$		$x_4 = 0.95614689$	$x_4 = 0.95614718$
$x_5 = 0.56541399$	$x_5 = 0.52437216$	$x_5 = 0.51560414$		$x_5 = 0.16772007$	$x_5 = 0.16771899$
$x_6 = 0.56515831$	$x_6 = 0.79809919$	$x_6 = 0.79922083$		$x_6 = 0.55836020$	$x_6 = 0.55835733$
$x_7 = 0.56523510$	$x_7 = 0.51563799$	$x_7 = 0.51349406$		$x_7 = 0.98637635$	$x_7 = 0.98637769$
$x_8 = 0.56521207$	$x_8 = 0.79921745$	$x_8 = 0.79941731$		$x_8 = 0.05375220$	$x_8 = 0.05374697$
$x_9 = 0.56521898$	$x_9 = 0.51350053$	$x_9 = 0.51311767$		$x_9 = 0.20345161$	$x_9 = 0.20343294$
$x_{10} = 0.56521690$	$x_{10} = 0.79941675$	$x_{10} = 0.79944937$		$x_{10} = 0.64823620$	$x_{10} = 0.64819192$
$x_{11} = 0.56521753$	$x_{11} = 0.51311874$	$x_{11} = 0.51305625$		$x_{11} = 0.91210412$	$x_{11} = 0.91215662$
$x_{12} = 0.56521734$	$x_{12} = 0.79944928$	$x_{12} = 0.79945451$		$x_{12} = 0.32068079$	$x_{12} = 0.32050770$
$x_{13} = 0.56521740$	$x_{13} = 0.51305642$	$x_{13} = 0.51304639$		$x_{13} = 0.87137848$	$x_{13} = 0.87113005$
$x_{14} = 0.56521738$	$x_{14} = 0.79945450$	$x_{14} = 0.79945533$		$x_{14} = 0.44831210$	$x_{14} = 0.44904993$
$x_{15} = 0.56521738$	$x_{15} = 0.51304642$	$x_{15} = 0.51304481$		$x_{15} = 0.98931345$	$x_{15} = 0.98961636$
$x_{16} = 0.56521738$	$x_{16} = 0.79945533$	$x_{16} = 0.79945547$		$x_{16} = 0.04228940$	$x_{16} = 0.04110327$
$x_{17} = 0.56521738$	$x_{17} = 0.51304481$	$x_{17} = 0.51304456$		$x_{17} = 0.16200405$	$x_{17} = 0.15765516$
$x_{18} = 0.56521738$	$x_{18} = 0.79945546$	$x_{18} = 0.79945549$		$x_{18} = 0.54303495$	$x_{18} = 0.53120005$
$x_{19} = 0.56521738$	$x_{19} = 0.51304456$	$x_{19} = 0.51304452$		$x_{19} = 0.99259197$	$x_{19} = 0.99610623$
$x_{20} = 0.56521738$	$x_{20} = 0.79945549$	$x_{20} = 0.79945549$		$x_{20} = 0.02941259$	$x_{20} = 0.01551444$
$x_{21} = 0.56521738$	$x_{21} = 0.51304452$	$x_{21} = 0.51304451$		$x_{21} = 0.11418998$	$x_{21} = 0.06109498$
$x_{22} = 0.56521738$	$x_{22} = 0.79945549$	$x_{22} = 0.79945549$		$x_{22} = 0.40460251$	$x_{22} = 0.22944952$
$x_{23} = 0.56521738$	$x_{23} = 0.51304451$	$x_{23} = 0.51304451$		$x_{23} = 0.96359727$	$x_{23} = 0.70720976$
$x_{24} = 0.56521738$	$x_{24} = 0.79945549$	$x_{24} = 0.79945549$		$x_{24} = 0.14031027$	$x_{24} = 0.82825647$

Fig. 3.4 The population size values x_0 to x_{24} for different values of maximum fertility r: (a) $r = 2.3$; (b) $r = 3.2$; (c) $r = 4$. In each case, we start with two different values for x. In (a) and (b), different initial conditions tend to the same values. In the latter case, a change in the eighth digit leads, after 20 steps, to a totally different outcome. If we had changed the sixteenth decimal, the error would take twice as many steps to explode.

equilibrium—always the same value, no matter where we start from. If r is slightly larger than 3, however, the x_n alternate (after a short initial phase) between two values, always the same, no matter again where we started from. So if the fertility r of our model population crosses the threshold value 3, the equilibrium point *splits* into a cycle consisting of two points. This so-called *bifurcation* is well understood.

But what happens for still larger values of r? At $r = 3.47$ the two-cycle is replaced by a four-cycle, which at $r = 3.54$ gives way to an eight-cycle, and so on. Hence, increasing fertility leads to a *cascade* of bifurcations, always doubling the former period. The bifurcation values follow each other ever more closely, and accumulate at $r = 3.5700 \ldots$ After that, things definitely get weird. Some r-values lead again to a well-specified periodic regime; some others, to aperiodic oscillations which seem completely irregular. The motions become unpredictable: the smallest change in r or x_0 may cause a totally different behaviour (Fig. 3.4).

So the simplest conceivable rule for population growth offers an amazingly complex, fascinating behaviour. Take for instance the value $r = 4$. For any sequence of Heads and Tails, finite or infinite, one can find a starting value x_0 such that the population sizes take on values above 0.5 (for Heads) and below 0.5 (for Tails) in exactly the specified order. If the twelfth toss came up Heads, for instance, then x_{12} would be larger than 0.5. Hence the iteration of the strictly determined computation leading from x to y yields, for a suitable starting value, exactly the random results obtained by tossing a coin. For close-by starting values, one obtains sequences which at first behave just the same, but later evolve on their own, quite independently of one another.

Long before May, ecologists had observed that population numbers were apt to oscillate irregularly, without ever settling down to a definite period. They used to explain this away by invoking erratic oscillations of the environment, due to weather, maybe, or—since the jars of flies were kept at constant temperature—to unpredictable outbreaks of disease. And this may well be right. But May's model shows that it doesn't need any complicating factors to cause complicated behaviour. The very simplest type of self-regulation can already account for it.

Oddly enough, not all of this was new. Several mathematical ecologists had, from the early 1950s onwards, stumbled upon chaos in their models. They did their best to hide it.* But in 1976, May's paper on 'Simple mathematical models with very complicated dynamics' brought chaos into broad daylight.*

And chaos took off. The time was ripe for it. May's message fitted in nicely with a handful of intellectual sparks which, while not altering the scientific view of the world, have changed our perception of it. The new-born 'chaotology' has forced us to revise the meaning of that central tool of science, *computation*, and to back-pedal on that age-old aim, *prediction*.

The French astronomer and mathematician Laplace is reported to have said 'Sire, I don't need this hypothesis', when Napoleon asked him about the role of God in his (Laplace's) celestial mechanics. But for explaining things to the lay mind, Laplace made use of a *demon* who in due course became the emblem of the deterministic world-view. This being was provided with knowledge of the exact state of the universe at some given instant in time. Just one glimpse; but enough for a demon. Since the laws of physics determine all the future and past, the demon can compute the exact state of the world for any other time it cares to know about.

Cuvier (a contemporary of Laplace) was reputedly able to deduce from a single bone the shape of a dinosaur. Our demon can likewise deduce from one single still the whole film of the universe, from opening shot to fade-out.

Laplace belonged to the illustrious band of mathematicians who had refined and extended Newton's method for determining the motion of the planets. He therefore laboured under no illusion as to the difficulties of such computations. Newton had been able to analyse completely the model consisting of the Sun and *one* planet. His solution of this two-body problem can be explained to any sixth-former today. But so far, the three-body problem (Sun, Moon, and Earth, for instance) is unsolved, in spite of the most valiant efforts. One doesn't even know whether the Moon will tumble down; nor can one rule out that the Earth will, some day, escape the pull of the Sun. One can only compute approximate values for the positions and velocities of celestial bodies. The precision of this approximation can be increased, but at soaring costs of computer time. Much more importantly still, the calculations depend upon the initial conditions—positions and velocities at a given moment. And short of being a demon, one does not know their precise values. One is saddled right from the start with a small error, whose size will increase at every step of the calculation, in a chain reaction which eventually will get out of control. Eclipses of the moon, for instance, can be predicted hundreds and thousands of years in advance, but not for millions of years (which, by astronomical standards, is a very short time).* The billionfold increase in the speed of calculations due to computer technology has improved things, to be sure, but less than one might expect.

In the early 1950s, John von Neumann had hoped to forecast weather, by means of successors of the ENIAC computer, for many weeks and months ahead. He was so confident in having solved the problem, in fact, that he was all set to grapple with the next task ahead—weather control. But 10 years later, the long-expected breakthrough in forecasting had failed to materialize, and the American meteorologist Edward Lorenz was trying to find out why. He toyed on his computer with a radically simplified model of the atmosphere and stumbled, quite unintentionally, upon the fact that an initial difference eight digits behind the decimal point led to a totally different outcome for his calculations. This is the famous *butterfly effect*: the flap of a butterfly's wing in Brazil could cause, some weeks later, a tornado in Texas.*

The keys to paradigms

Together with May's abstract insect populations, Lorenz's meteorological artefact played a pivotal role in the evolution of science in the following decades. In astronomy, hydrodynamics, physiology, mechanics, chemistry, and economics, chaotic trajectories sprang up like flowers after a rainfall in the desert. Everywhere, one found motion oscillating in nervous outbursts and reacting with utmost touchiness to the slightest change in initial conditions. This motion settled on *strange attractors*, where nearby points were blown apart and widely separated regions brought together, in a relentless stretching and folding similar to the kneading of a piece of dough.

Such a chaos is *deterministic*. Chance has no say in such a process, in principle at least. The dynamical system allows only *one* future to emerge from any beginning. Such a system can be completely docile, as in the case of a pendulum or of a weight oscillating on a spring. In these examples, a small change in initial conditions will not get amplified. This yields nice exercises (always the same) to furnish textbooks with, and suggests to the students a world which is easily mastered; but this world does not extend very far beyond the covers of the books. Mathematicians knew of other examples with incurably complex behaviour. But they had problems in putting this across. It took pocket calculators to check painlessly that a few iterations of a simple computational step may yield a completely different outcome if a digit is modified eight places behind the decimal point. The computable need not be predictable. Any small seed of uncertainty can grow into an avalanche of error.

Of course, we have always been living with uncertainty. We have even known how to produce it by deterministic means. The toss of a coin, for instance, is completely specified (at least in classical mechanics) by its 'initial condition': spin, angle of its throw, elasticity of the table top, and so on. But the initial conditions leading to Heads are so intimately interwoven with those leading to Tails that it is impossible to predict which side of the coin will turn up. No one wondered about the uncertainty of the outcome, however, since the equations describing the throw were horribly complicated. In contrast, the equations of May and of Lorenz are of a very simple form, mathematical *haiku* which show that chaos is not far away.

As ever casino owner knows, one can rely on uncertainty. One cannot predict the outcome of a single throw, but one can predict the outcome of a large number of throws. In exactly the same statistical way, chaos is predictable. It is not possible to know the population size 12 generations from now, but it is possible to predict the average for the next 100 000 generations. Actually, this was discovered by John von Neumann and Stanislas Ulam way back in the late 1940s, when they performed some of their first experiments with computers. One cannot say when the iterates of $y = 4x(1-x)$ will lie in the interval between 0.1 and 0.2, for example, but they will do so, in the mean, in 12 per cent of all generations. It is precisely because of the instability of the motion, in fact, that its average can be so easily tamed; just as a baker's kneading would quickly mix a drop of dye uniformly throughout the dough.

So if you don't want to use dice, you can use a PC. In fact, the random number generators are all based on deterministic chaos. The fish on the planet Wa-Tor, for instance, performing their random motion which leads them with the same odds to any vacant neighbourhood, are strictly governed by arithmetical rules.

There is more order in chaos than just of the statistical type. The story of Mitchell Feigenbaum discovering his magical number has been told so often and so well that it belongs by now to the folklore of science.* True to our time, it is about a scientist kicking his computer, and getting a kick in return. The kick being, in this case, that the cascade of period doubling which leads to chaos is universal: that it always looks the same, no matter what single-humped curve one takes. The parabola was just the first example to spring to mind. Feigenbaum understood that all single-humped feedback devices show the same 'contraction rate' 4.66920166 . . . in their bifurcation sequence, and a few years later, mathematicians duly proved him right.

The cascade had turned into a flood by then. From population regulation, it had spilled into many other fields.* It was found in turbulence, for instance, this scourge of physicists; in the dripping of a tap; in the oscillations of a double pendulum; in cardiac arrythmias; in the outbreak of measles; in the motions of an eye-ball; and so on. A door had been opened into a new world. The old paradigm of a clockwork universe drifted off.

The unconscious has its role in science, no less than in the individual's mind. To set up and falsify hypotheses belongs to the well-regulated, conscious part of scientific enterprise. But many intellectual straightjackets and blind spots constrain research in ways which can be noticed only after being overcome. This background of pervasive prejudice is the true frontier of science.

Besides the subliminal influences from religious and political ideologies, it is often the insights from one science which mould the expectations and understandings in another. Thus Darwin was steeped in the doctrine of Malthus on overcrowding and shortage of living space, as well as in the way of thinking of the geologist Lyell, who explained the gigantic changes of the Earth's past by the persistent action of forces still operating today. Technological progress was a similar source of inspiration. The concepts of energy and entropy came from a close acquaintance with steam engines, and the view of heredity as information transfer could only have taken hold in a century shaped by advances in communication.

Not all idea transplants were successful, of course. The *physics envy* of biologists has sometimes led them astray. Thus chronometers were hardly ticking yet when the living organism was compared to a clockwork. As soon as physicists tinkered with their first batteries, the 'living spark' was purported to be electrical. Magnetism was immediately pressed into service as 'animal magnetism'. And the very first computers, huge chests filled with vacuum-tubes, were hailed as the key for understanding the brain. All this left more lasting traces in science fiction than in biology. Sometimes, however, a spark set a powder keg off—Schrödinger's idea of an aperiodic crystal, for instance, gave a big boost to molecular genetics. But no matter whether it turned out to be humbug or inspiration, the impulse always seemed to jump from physics to biology, according to a God-given hierarchy; until Robert May (himself a renegade from physics) could gleefully report that it worked the other way too.

Niche pickings

The model which had launched this revolution described population growth for *one single* species. But there are no isolated species outside of the laboratory. Even astronauts are not all by themselves, since they carry a rich flora in their intestines. The circuitous ways in which one species can affect others form a *leitmotiv* of ecology. Darwin revelled in such unexpected links. 'It is quite credible', he wrote, 'that the presence of a feline animal in large numbers might determine, through the intervention first of mice and then of bees, the frequency of certain flowers!'*—And, in an intriguing preview of deterministic chaos—'Throw up a handful of feathers, and all must fall to the ground according to definite laws; but how simple is this problem compared to the action and reaction of the innumerable plants and animals.'*

To get a clearer picture of such interactions, let us eliminate the cause of the instability so far—the feedback's time-delay. In its absence, an isolated population settles down tamely to a steady state delimited by the environment's carrying capacity. This is simply due to competition within the population. Every supplementary member impedes further growth. But what happens if *two* distinct populations compete with each other?

Let us assume that one population has been living all by itself in restful equilibrium, and that another invades. The newcomers will, at first, be present in a minority only. The environment's capacity is already used up by the indigenous population. So all depends on whether or not the newcomers can displace some original inhabitants. If not, the invaders will be ousted forthwith. If yes, they will multiply. This allows three possible scenarios.*

1. It could be that each species can invade the other. They will then end up by *coexisting* in the same habitat.

2. It could be that one species can be invaded, but cannot invade in turn. In this case, it will obviously be *dominated* and doomed to disappear.
3. It could be that neither species can invade the other. This is the case of *bi-stability*. Whichever species happened to intrude would be vanquished— much as Manchester United is more likely to win on its own turf against Liverpool, and vice versa.

Both in the case of dominance and of bi-stability, one species outcompetes the other. But in bi-stability, who is going to win depends on the initial numbers; the minority group will be eliminated. In dominance, one species is better than the other and will drive it to extinction, no matter where it started from.

In the case of bi-stability and of coexistence, both competing species can live side by side in equilibrium, with their death rates balancing their birth rates exactly. But if a fluctuation creates a small surplus of one species (as it inevitably will), then this imbalance will be compensated for in the case of coexistence, while it will grow and lead to the extinction of one species in the case of bi-stability.

If one looks for biological examples of two species in competition, one finds that they are as infrequent as the single species living all by itself. There is always, in nature, a plethora of busybodies meddling with what one would like to model as two rival populations in an intimate 'bête à bête'. We must therefore look for laboratory populations. So, it's back to the bottle again, or more precisely to the glass tube, for the first experimental studies on competition were done by G. F. Gause on protozoans, in Moscow during the 1930s.*

Two populations compete if every increase in one lowers the growth rate of the other. This can happen in various ways: by using up resources (food, space, or light) or by secreting toxins, for example. Gause was able to find examples for each of the three types of competitive outcome. But in setting up his battles in the test tubes, he learned something more: whenever scarcity in *one single* resource was the limiting factor, one species was dominant. Coexistence and bi-stability occurred only if *several* factors played a role. This became known in more general terms as the *competitive exclusion principle*: there cannot be more species than resources around.

Enunciating principles in ecology is a risky business, and the exclusion principle promptly came under fire. Mathematicians proved that it would only hold under rather restrictive assumptions on how resources affect the growth rates; and biologists pointed out that it was not easy to delimit resources. How many resources does a spruce tree, for instance, offer to warblers? Once it is agreed that the different parts of a plant, or its different stages of growth, form so many different resources, it is hard to see where to stop.

Nevertheless, Gause's principle proved singularly fruitful; for it led ecologists like Evelyn Hutchinson and Robert MacArthur to the notion of the *niche*.* Ecological niches are not simply places: they are jobs, or more precisely, ways of obtaining resources. MacArthur was able to show, for instance, that different kinds of warblers worked on different parts of the spruce tree. Should the principle be that each niche can sustain one species only? But for this to be meaningful, one would have to know the boundaries of the niches. In other words, how similar can two species be, and still live together? This leads straight to some central (and wide open) questions of evolutionary ecology.

Jolts of fortune

Just as in celestial mechanics, where the two-body problem is easy and the three-body problem essentially unsolved, so the competition of two species is simple and that of three quite intractable. One of the reasons is the possible occurrence of cycles which eventually get stuck, but unpredictably so.

Here is a thought experiment. Let us imagine three competing species such that the first dominates the second, which dominates the third, which in turn dominates the first. If this strikes you as an unlikely set-up, I agree. I know of three tennis pros in such a cyclic rank ordering, and even of three laboratory populations,* but not of three competing species in nature. It is interesting, nevertheless, to figure out what would happen, taking to heart an exhortation from J. B. S. Haldane's *Possible worlds*:* in order to gain a proper perspective of our own world, we should try to conceive alternative realities, even if they seem fantastic at first. R. A. Fisher said much the same thing when he proposed studying species with three sexes.

Knowing the outcomes of the pairwise interactions in our ecological Stone–Scissors–Paper game is not enough to predict the outcome of the full three-species system. It will spiral, that much is clear: this tendency is built into the cyclic ordering. But depending on the rates of invasion and extinction, two radically different results can emerge.* In one case, the three species will settle down to a stable coexistence at well-determined equilibrium values or regular cycles, safely cushioned against perturbations. In the other case, they lash themselves into more and more violent outbreaks. The Stones seem, at one time, to get the upper hand; but since they are dominated by the Papers, they have to give way; the Papers take over, but this, after a time, causes an upswing in the populations of Scissors; the Scissors' hegemony will be very pronounced for a while; but suddenly, without an exterior cue, the Stones, which had been on the verge of extinction, rally round and come back, dominating the community much more than they did in the previous round. The next upturn takes a very long time to come, but come it will: the Papers drive the Stones against the ropes. And so it keeps going, in an endless cycle of revolutions. It takes about the same time, always, for an upcoming population to grow from, say, 5 per cent to 95 per cent of the total: these upheavals are short and sharp. But they become more and more rare (in fact, the time spans between consecutive shifts grow exponentially), and they get more and more violent (the population sizes of the current minorities shrink from one cycle to the next). After a while, observers will think that the struggle has reached a decision: for a very, very long time, they can make out one population only. But the mechanism for the next revolution keeps working underground, hidden in a minority too small to be detected, and in due time will cause another sudden reversal of fortune.

So much for the model. Out there in real life, the minorities cannot become too small; if their population number drops below one (or two, if reproduction is sexual), then the game is over. Among the remaining two populations, the dominant will emerge as the sole survivor. Thus the cyclic evolution is ultimately doomed to stop for good. The same holds, by the way, for the computer simulations. At some time, one of the numbers will be so small that the computer rounds it up to 0.00000 . . . and that's it. But the ultimate winner may depend in a very delicate way on initial values, so that we have no way of predicting the outcome. Moreover, the final equilibrium is always good for a surprise: indeed, a tiny minority of the 'next' species introduced, for instance, through migration from another location, will unstoppably grow, and cause another jolt.

This Stone–Scissors–Paper story is meant as a fable. Admittedly, no one has ever found such a jerky roundabout in nature. It is hard to imagine how it could ever be observed, in fact. But this does not exclude the possibility that some apparently stable aspects of a present-day ecosystem have been caused by mechanisms of this sort, and are all set for sudden change. A food web does not consist of three, but rather of dozens or thousands of species interacting via rivalry, mutual help, and exploitation. With so many feedbacks, the system is perhaps only provisionally in a steady state, and ready to react to a small perturbation by a sudden switch in an entirely unforeseeable direction. Some of the populations will crash, others explode, possibly only for a brief burst, triggering new transitions in turn, until the system winds up in a totally new rearrangement of species distributions, seemingly stable but possibly quivering, once again, with the potentialities for another upset.

A switch at the top

In nature, food chains consist usually of three or four links, although systems with 10 links have been observed.* Thus many predators are subject to predation in turn. Simple prey–predator systems will oscillate, as we know, if the predator is good at its job. A system consisting of prey, predator, and superpredator should, one feels, oscillate all the more, at least if both predator and superpredator are efficient enough. Recently, several ecologists have modelled such three-layered feedback device,* and come up with some beautiful cycles indeed.

Here is an example: we start with almost no superpredators around, and violent oscillations of predator and prey. Slowly but surely the few super-predators multiply. As their number increases, the prey–predator cycle dampens down. The superpredators reduce the efficiency of the predators. For this reason, the oscillations between prey and predator gradually give way to a stable steady state. But the superpredators will still increase at a slow pace. The strain on predators grows to a breaking point: suddenly, they give way and crash to a very low level. But now, rescue arrives from two sides at once. On the one hand, the prey level rapidly zooms up, since it is almost freed from the predator's load. On the other hand, the superpredators are on the wane, having nothing left to feed upon. Still the prey needs time to recoup its losses: by the end of this time, the superpredator will be on the verge of extinction. So when the predator comes back, it is fully efficient again, and can return to its violent spinning around with the prey.

To summarize: the superpredator slowly waxes and wanes; the predator starts out with a sudden burst of violent oscillations, tames down into a smooth decline, suddenly crashes, and is slow to take off again; but when it does, it will repeat its crazy wiggles. Such a manic-depressive succession of wild outbursts and lethargic inanimation can be observed in some population curves: in fact, the intermingling of high- and low-frequency oscillations seems to be a familiar feature in nature. It is too soon to decide whether this sometimes reflects a prey–predator–superpredator scenario or not. But the Italian ecologist Sergio Rinaldi points out that in some lakes with high eutrophication, the data on the three-layered system consisting of phyto-plankton, zooplankton, and fish show just the right type of oscillations.

Some unexpected properties of complex ecological webs have been uncovered not by theoretical modelling but by direct observations. The strange role of the top predator in stabilizing communities is a case in point. Top predators are those which are not preyed upon, the ultimate exploiters of the ecosystem. They are usually scarce, and therefore liable to become extinct. Such an extinction must be a bonus to the rest of the community, one should think. But this is not the case at all. It was observed time and again that the disappearance of a top predator led to an impoverishment of the community's diversity. Matters were put to an experimental test by the ecologist Robert Paine, who removed the top predator—a starfish—from an intertidal community consisting of sixteen species.* Two years later, only eight species remained. Apparently, we have to think of a food web as a kind of vaulted arch, with the top predator as its keystone. The arch collapses when the keystone is removed.

This strange result reminds one of some revolutions in history, where the dispersal or execution of the most powerful class (of so-called exploiters) led to a drastic reduction of the diversity of the exploited.

How do the predators manage to increase the diversity of their prey? Essentially, so it seems, by interfering with the results of competition between prey. If one prey species dominates another, for instance, predators can specialize upon it and curtail it to such a point that the other species, although it may also get preyed upon, can hold its place in the habitat. The population sizes may spin chaotically without ever coming close to zero.* If two prey species are engaged in a *bi-stable* competition, however, then one predator doesn't seem to be enough to mediate a permanent coexistence.* *Two* species of predators can do it, however, if each is properly specialized. So can one predator, after all, provided it is good at playing two parts: if it can switch, whenever one prey gets scarce, to preferentially hunting the other. Exploiters are notoriously opportunistic. A highly evolved predator is likely to be brainy enough to adapt its search image and its hunting technique to whatever currently prevails on the market. Such a *switching predator* may keep a community balanced by checking the growth of whichever prey gets the upper hand.*

That lynxes can switch has been documented by A. T. Bergerud's study of Newfoundland's ecology to which we have alluded already.* On this huge island, there are only 14 species of mammals, of which nine are carnivorous—an uncommonly high proportion. In 1860, snowshoe hares were introduced by humans. Their population exploded in the next fifty years, and then settled to an apparently self-driven cycle with a period of roughly ten years. Enter the lynx, which had formerly been absent, or at best reduced to a marginal role. The lynx cashed in on the snowshoe hare bonanza, and multiplied pro-digiously. This way, it locked onto the snowshoe hare's cycle. Whenever hare populations fell, the lynx took a dip. This dip, however, was less pronounced than might have been expected, because the lynx proved flexible enough to switch in case of need to a diet of caribou calves. Caribou numbers fell, but by the time lynxes began to suffer from their decline, snowshoe hares were back at peak level, enabling the lynx to switch back. The caribou numbers recovered, and all was ready for another cycle.

A native resident, the arctic hare, was depleted by the immigrant lynx. Had the lynxes been reduced to a diet of arctic hare only, they would have crashed as a consequence, and their prey could have recovered. As it was, the lynx managed, by switching, to persist at high density, forcing the arctic hare to stay trapped in its predatory pit.

Power to the parasites

Humans are predators, and very good at switching, by the way. It may be for this reason that we frequently take a keen interest in other predators. We all feel that the pace quickens when predators are around. But only recently have ecologists begun to appreciate the power of parasites, which live, like predators, by exploiting other species and are just as good, or even better, at driving population numbers up and down. This long neglect is all the more astonishing as Lotka's very first model had dealt with the cycle of mosquitoes and humans in transmitting malaria.*

In retrospect, however, it seems obvious. Disease can kill. Parasites act upon their host populations like predators upon their prey. They can cause fluctuations. Volterra's principle holds again: if one population exploits another and if the growth rates of both are reduced, this leads to fewer exploiters and more of the exploited. Actually, Darwin had already derived this principle without using any mathematics at all. 'If not one head of game were shot during the next twenty years', he wrote, 'and, at the same time, no vermin were destroyed, there would, in all probability, be less game than at present.' (This because the game—partridges, grouse, and hares—fall *prey* to the vermin.)

Parasites can drive oscillations.* They can keep hosts in an evolutionary pit. They can provide for the coexistence of competing host species. They cannot switch their foraging strategy in the way of a flexible predator, but some parasites can adapt in an amazingly short time to ecological change by evolving genetic variants. And parasites, of course, may be parasitized in turn. 'Big fleas have little fleas, and little fleas have lesser fleas', according to Jonathan Swift.

This early science fiction writer would have been fascinated by the recent accounts from *Tierra*, a cybernetic universe devised by the ecologist Thomas Ray* which has been hailed as the 'most sophisticated artificial-life program yet developed'.* In this artificial ecosystem contained within the memory-space of a computer, short self-replicating programs compete for processing time. In one such experiment, Ray started out with 20 types of competing programs, of which only eight were left at the end of the run. Then, Ray repeated the experiment, adding one program which was parasitic in the sense that it borrowed part of the instructions of other programs. In this second run, 16 of the original 20 types of host programs survived—an amazingly close analogy to Paine's real-life experiments with intertidal 'wetware', where the keystone predator kept any of the prey species from driving others to extinction.

The interest of theoretical biologists in parasites has boomed within the last few years. This did not always cause unmitigated delight to parasito-logists, who point out—quite rightly—that a host means much more to the parasite than just a source of food. It is frequently also a shelter, and a vehicle for propagation, a kind of second-order body in a sense. Whenever the parasite depends, for transmission, on a host in full working order, then it is under strong selective pressure to treat its host well, in the sense of evolving into a benign rather than a virulent strain:* such a parasite can eventually turn from exploiter into mutualist. These parasites will not lash their host population into violent oscillations.

Eggs and islands

Model 'ecosystems' with three or four species are barely tractable. The mind boggles at the complications that real food webs come up with. And yet, for all their unpredictability, they display patterns in time and space which are of intriguing regularity. This fact is so striking that many naturalists, to this day, tend to see ecosystems as super-organisms, responding to perturbations with the coherence of an individual body recovering from injury. This is naïve, of course. Ecosystems are not built by selection in the same way as organisms are. Their regularities must be caused, like those of sand-dunes or waves, by external forces and constraints.

There is a widespread feeling that the stability of ecosystems is due to their very complexity, but mathematical models have failed to support this. However, both simulations and experiments tend to show that the patchiness of the environment frequently ensures the persistence of otherwise unstable eco-communities.* If some host–parasite cycles, for instance, lead to local extinction in one patch, then dispersal from a neighbouring patch will recolonize it. It is a way of hedging bets. The spatial patterns can be just as chaotic as the fluctuations in time; but their overall effect is to cushion the extremes and dampen the instabilities.

In the first decades of the century, the recolonization of patches provided the budding science of ecology with its first great theme: the orderly succession of plant communities.* If farmland gained upon a forest is subsequently left to go wild, then it is invariably taken over by annual weeds, which later give way to perennial weeds. These in turn are displaced by bushes, which yield, in due time, to trees. To some of the early ecologists, it looked as if the pioneer stages were there for the purpose of (literally) preparing the ground for the forest. In the harsher light of Darwinian live and let die, this cooperation of species for the good of the eco-community turns out to be a fond dream, of course. The early assumptions that the final, climactic phase was richest in species and best at converting the energy from the sun does not withstand closer scrutiny either. But succession remains a fact.

Its stages are explained by the economic strategies of different species. Some plants turn all their resources into mass-producing seeds as quickly as possible; others spend more in keeping alive for a while. There are all kinds of intermediate recipes, of course, based on hedging the bets: but there is necessarily a trade-off between longevity and fecundity, between cashing in and salting away.

If a habitat opens up (as, for instance, when a farm closes down), the opportunistic species move in fast. They take the money and run, without bothering to put up much of a fight when the serious competition starts showing up. They cannot be asked to hold their ground against plants which prepare for the future by stockpiling reserves or by building large roots or even by deploying armaments (for instance, growing stems to over-shadow the rivals). The quick-buck gambit has to wait for opportunities to arise; it thrives in environments which are transient and patchy. A tight-fisted policy of hoarding up, on the other hand, is appropriate for holding one's own in stable environments where all available niches are filled to saturation. This simple principle goes a long way towards explaining ecological succession.

Islands are the ready-made testing grounds for ideas on ecosystems. Much as a traveller will, from the size and location of a town, know more or less what to expect—number of restaurants, cinemas, shops, and mechanics—so ecologists will form definite ideas about an island's eco-community, just from a look at the map. They would not buy the story of King Kong's island, which is far too small to provide a habitat for the many species of huge beasts wrestling over the heroine.

Ecologists can even make fairly precise predictions. In a rather revolting experiment, some biologists fumigated a few tiny mangrove islets,* destroying all animal life, and watched their recolonization: the species which then appeared were in general quite different from what had been there before, but the total numbers of species returned to the previous levels, much as expected.

Some natural experiments confirm this: the recolonization of volcanic islands, for instance Krakatoa. Another striking example is provided by what Stephen Jay Gould has called the worst ecological catastrophe in the last few million years:* the formation of a land link between the continents of North and South America. No family of mammals had been common to both continents: there were some 25 families of mammals in North America and some 30 in the South. The rise of the land bridge was followed by a period of intense turmoil as species migrated to new grounds and competed against new rivals. Many species were driven to extinction (including lots of southern marsupials). But when the dust settled, the count was, again, 25 families in North America and 30 in the South.* Some actors had been exchanged for others, but the roles in what Evelyn Hutchinson has called the 'ecological theatre' remained the same as before.

The tur-balance of nature

Mathematical models are still far from being able to account for such regularities, and to allow, as MacArthur had hoped, predictions of the form 'for organisms of type A, in environments of structure B, such and such relations will hold.'* So far, the main contribution of mathematics to ecology has been to expand our sense of what is possible. The turbulent behaviour of extremely oversimplified 'toy' populations has served as an eye-opener to biologists all too accustomed to assume that ecosystems have an innate tendency to develop towards some steady equilibrium. The formerly wide-spread feeling that nature abhors instability has abated by now. But there is no denying that ecosystems are organized along predictable overall patterns, in spite of being filled with erratic components. The particular populations are there by historical accident, and subject to odd vagaries; whole systems are, nevertheless, amazingly well behaved. They are like the weather, which cannot be predicted for three weeks in advance, and yet will fit in with a climate which holds steady for centuries. The butterfly effect cannot alter averages. Statistical regularities persist.

Chaotic motion is surprising in the short run, but keeps repeating, although in an unpredictable fashion, over and over again. This blending of surprise and repetition is frequently found in the evolution of ecosystems, but it is not restricted to the ecological scene.

The tension about the uncertain outcome is but one aspect of play; another is the reassuring ritual of recurrence. Some gift shops nowadays sell little gadgets displaying deterministic chaos. You may have one of these devices on your desk. They consist of a few shiny weights, rigid rods, and excentric loops mounted on almost frictionless bearings. Some are driven by electricity, but others work by purely mechanical means. While not exactly perpetual, their motion hardly ever stops. The smallest vibrations from the traffic in the streets below, the tiniest currents of air keep them going. In spite of having only two or three axes of rotation, their movement is chaotic. If you watch one of the small metal weights, you will see it swing to and fro, pick up momentum, tumble twice, do a quick turnabout, and almost come to a stop, then burst into a frenzy of activity, change course, come to a halt, work it all up anew, tumble six times in another direction, and so on. A few decades ago, no one would have dreamed of constructing such toys. And yet, the mechanical know-how was there: in fact, clock-makers of the seventeenth century could have produced them. But they preferred to build chrono-meters instead, and their customers preferred it too. Motion was meant at that time to be regular, stately, and periodic. The clock symbolized the world. Our generation takes another view. We delight in erratic mechanics and the loss of balance.

But if we keep watching one of those little chaotic toys, we find that we soon get wise to its tricks. We cannot say whether its next somersault will be forward or backward, but get a distinct feeling for the *style* of the motion. It is as plainly recognizable as a particular handwriting, and as difficult to pin down. And it reveals an elusive balance behind the restless twists and spinnings of the toy. An ecologist on the look-out for the balance of nature is after similar game.

Random drift and chain reactions

4

Random drift and chain reactions
A random walk through the gambling hall

. . . Not yet exempt, but ruling them as slaves,
From Chance, and Death and Mutability. . .
Shelley

Name dropping

Children traditionally inherit their father's name, although it is sometimes not as well ascertained as their mother's. By this strange *nomenklatura*, the male branch gains a surplus of attention which is biologically quite unfounded. The extinction of a family name is often equated with the extinction of the family itself. Even today, many comely flocks of daughters testify to their father's tireless desire for a male heir.

Mathematical studies on the disappearance of rare family names date back to the nineteenth century. Statisticians rigged up a properly simplified model—later called a *branching process*—and derived from it that every family name would inexorably have to vanish sooner or later.* This was a heraldic counterpart, so to speak, to the *Wärmetod* (heat death) which the contemporary physicists prophesied for our passing world—a disconsolate perspective for a century hooked on progress. Both results turned out to be wrong, by the way. The *Wärmetod* is out, like the same physicists' *ether*—today's cosmologists have other fates in store for our world. And the apparently inescapable extinction of family names was based on a simple miscalculation; but it took a long time to discover the error.

Since it is usually only sons who ensure the survival of a family name, it is enough to keep track of male descendants in our calculations. Now a man, it turns out, has with a probability of 48 per cent *no* son; he has *one* son with a probability of 21 per cent; *two* sons with 12 per cent, *three* sons with 8 per cent, etc.* Each son can have zero, one, two, three, or more male offspring, who in their turn can have sons, etc. We will assume that the probability to have so and so many sons is always the same; and that the number of sons of any one man is independent of the number of sons sired by others. Both assumptions are only approximately valid; average fertility changes with place and time (here we used values from the USA in the 1920s), brothers tend to have similar family sizes, and so on. We shall neglect all this for simplicity's sake.

The male line goes extinct in the very first generation, if no son is born; or in the second generation, if no son fathers a son; or in the third generation, etc. (Fig. 4.1). But it could also happen that the line *never* reaches extinction. What is the probability of this event? The original, and erroneous answer was: *zero*. We shall later return to the reason for this mistake. The correct formula, however, yields a value of 18 per cent. This is heady news for those perched on genealogical trees; there is a sporting chance that their name will last forever (at least if the end of the world is omitted from consideration).

Branch hopping

The destiny of old family names need not affect us profoundly; but the results from the branching model have wider-ranging applications. Bacteria, for

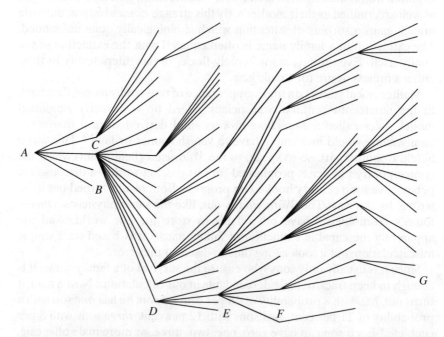

Fig. 4.1 A branching process in the Habsburg family. *A* is Franz von Lothringen, the husband of Empress Maria Theresa (1740–80). He was no Habsburg himself, but took on the name when the male Habsburg line died out in 1740. A similar extinction seems very unlikely nowadays. *B* is Josef II (1780–90), *C* is Leopold II (1790–92), *D* is Franz (1792–1835), *E* is Ferdinand (1835–48), *F* is Franz Josef (1848–1916), and *G* is Karl (1916–18). The dates are those of their reigns.

instance, do not replicate like the scions of noble houses, but their propagation can be described in much the same terms: offspring have offspring having offspring in turn. And even inanimate nature works this way. If a free-ranging neutron splits an atom of uranium, it releases more neutrons, which may in turn split more atoms, etc. Every *chain reaction* follows this pattern. Whether such a reaction dies off or persists is nothing but the problem of the extinction of family names again. So let us take a closer look at branching processes.*

We have all been told the story of the chess player invited by some generous sultan to name a wish. The chess player asked for one grain of wheat on the first square of the chessboard, two grains on the second square, four grains on the third, then eight, sixteen, thirty-two ... No sultan could ever fulfil such a wish; not even American grain farmers could. Here we have the simplest branching process: each grain on one square gives rise to two grains on the next. Let us replace the grains by bacteria and the squares by successive generations: one bacterium splits into two, then four, then eight ... This yields exponential growth.

Now chance enters the play. Let us assume that a bacterium may die with some probability p instead of duplicating (Fig. 4.2). It becomes possible, then, that the population will vanish. The probability of this happening in

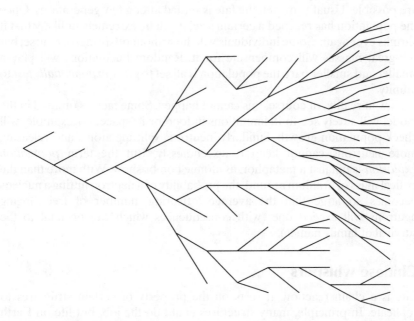

Fig. 4.2 A branching process: we assume that a bacterium can die with probability $p = \frac{1}{3}$; if not, it duplicates.

the first generation is p; if not, we have two bacteria. The probability that *both* subsequently die is only p^2; if this does not happen, then at least one will duplicate, maybe both. Theoretically, each generation is threatened by immediate extinction, but for a large population this risk is in fact extremely small. If the probability p for the death of a single bacterium is $\frac{1}{3}$, then the probability that such a population is reduced from seven to zero within one generation is less than $\frac{1}{2000}$. Nevertheless, a population is never *absolutely* safe. And by Murphy's law, it does not seem unreasonable to think that some time, in a future extending to infinity, such a disaster is bound to occur.

In fact, this need not always be the case. All depends on the average number of offspring produced by one individual. If this number is less than one, there is no escaping: it is 100 per cent certain, then, that the family will be wiped out. But if the average is larger than one, there is a sporting chance that extinction will be permanently avoided.

According to the statistics used above, a man has, on average, 1.31 sons; he may therefore entertain hopes of an unending sequence of male heirs. For the bacterium which dies with probability p and duplicates with probability $1 - p$, the average number of offspring is $2 - 2p$. This number is less than one if p exceeds 50 per cent, in which case the line is doomed for extinction. If p is less than 50 per cent, both eventual extinction and unlimited survival are possible. Usually, in fact, the fate is settled after a few generations. Once the population has reached a certain size, it will be extremely unlikely that it shrinks ever again. Some individuals will die without offspring, of course; but the great masses will compensate for it. Random fluctuations will play a smaller and smaller part: the population is all set to grow *exponentially fast* to infinity.

Now in reality, of course, this cannot happen. Some factors omitted in the model will surely see to it: limitations in food or in space, for example, will check population growth. Similarly, neutrons splitting atoms and releasing more neutrons cannot keep it up endlessly. But the term *population explosion* is not just a metaphor, as atomic bombs show. With more than the critical mass of uranium around, the probability of a neutron hitting a nucleus becomes so large that the average 'offspring' number of free-ranging neutrons will exceed one (with consequences which can be fatal to the survival of family names).

Chinese whispers

Life is a chain reaction. It rests on the property of certain structures to replicate. In principle, many structures could do the job; but life on Earth only uses a certain class of molecules called *polynucleotides*: strings of four kinds of chemical building blocks—G,A,C, and T for DNA molecules,

G,A,C, and U for RNA. For our purposes, polynucleotides are simply words written in a four-letter alphabet.

Copying such molecules will be subject to errors. The plainest are just misprints: one letter in place of another. These are called *point mutations*. There are many other possible errors: the copy can have fewer letters or more than the original, parts of the word can be permuted, or inverted, or mistakenly repeated. We shall forget about all this and stick to point mutations.

An erroneous copy can be copied in turn. Again, misprints can occur. But the probability that they lead back to the former text is negligibly small; they would have to occur at precisely the same place in the word. Therefore, we shall neglect such *back mutations*. It is obvious, then, that neither the mutants nor their descendants are copies of the original word any longer. For the task of propagating the original message, they may be viewed as irremediably lost.

Finally, we assume that a typographical error can occur with the same probability q at every place, irrespective of errors in other parts of the word. Again, this is not quite in line with what we know about the replication of such chain molecules. Some letters give rise less frequently to misprints than others; once an error is made, further errors become more likely; and there are *hot spots* in the message where point mutations are more frequent than usual. Again, we will not permit ourselves to be distracted by all these objections, valid as they are.

By taking thus our distance from reality, we can compute the probability that a word gets correctly copied. For this, obviously, every letter must be just right. With an error rate per letter of 10 per cent, for example, the probability that a word consisting of two letters gets correctly copied is 90 per cent of 90 per cent, that is, 81 per cent; for $N = 3$ letters, we get 73 per cent; for 10 letters, it is 35 per cent: 100 letters get correctly copied with a probability of 0.003 per cent, 1000 letters with 0.0000000000000000000000000000000000000002 per cent.

Now to return to the chain reaction: if we know how many copies are produced before the molecule dissociates (and thus dies), then we can estimate the average number of correct copies (or offspring). It has to be larger than one in order that the chain reaction has a chance to keep going. Hence, not too many mistakes are allowed. For the molecular message not to be doomed to extinction, the product of its length N with the probability q to misprint a single letter has to be below a given value. For every error rate q, this defines a maximal length. If the error rate q is halved, the number of letters can double.

In spite of all the simplifications used to derive this result, we do not seem to have strayed too far from reality. Nature confirms our crude guess, at least within an order of magnitude or two. There are several ways of copying polynucleotides, each with its own error rate q: they duly lead to different limits to the length of the chain molecule.

Errors in chains

The most slipshod replication has the message simply float around in a soup of letters. Since G pairs by chemical bonding with C, and A with U (or T), this can yield a complementary word, whose complement in turn will be a copy of the original. The error rate in such a soup is larger than 5 per cent and the maximal observed length has never exceeded some 15 letters or so. It is possible that under special conditions—by seasoning the soup with zinc ions or other mineral catalysts, for example—this length can reach 50 or 100. But one certainly cannot hope for more.

The very simplest viruses use a substantially improved method to copy their genetic information. They use a specific enzyme, a so-called *replicase*, which catalyses bonding quite efficiently. This replicase reduces the error rate by a factor of 100 or more (which already corresponds to the precision of a fairly good typist). The permissible word-length is accordingly at least 100 times longer. The genome of a Q_β-virus consists of 4500 letters—one or two printed pages.

Bacteria do better. On top of having their DNA copied by a replicase, they proof-read with the help of a supplementary mechanism, thereby checking the complementary strand of the double helix for eventual misprints. This quality control has to distinguish between the original and the complement, in order to know which letter to repair in case of mismatch. The *younger* of the two strands gets corrected. Compared with the virus, the precision is increased a hundred- or thousandfold. The acceptable word-length increases by the same amount. The DNA molecule of an *E. coli* bacterium, that most popular pet of geneticists, has a length of four million letters—a book with several volumes.

For higher organisms, whose cells are much more sophisticated, the *hi-fi* technology has taken a further stride ahead. On average only one in 10 billion letters gets misprinted. This allows for a further increase of the genome length by some three or four orders of magnitude. The genetic information of humans consists of 3 billion letters (several rows in a library). Other organisms, like some lizards or lung fishes, have up to forty times more DNA than we do.*

God, according to Goethe, has made sure that the trees cannot grow into heaven. With all due respect to God (and Goethe, needless to say), it may be added that it is the solidity of the wood, in fact, which limits the size of the trees. An oak, for instance, cannot be simply enlarged to scale by an arbitrary factor. If its height doubles, its weight increases eightfold, but the cross-section of its trunk, and hence its carrying capacity, only increases fourfold.

In a similar way, structure and material limit the size of land-bound mammals, for instance, or of insects (or, for that matter, of market halls and cathedrals). The limitation of the length of a genome by its error rate is of a comparable, *allometric* nature.*

Manfred Eigen, a Nobel prize-winning chemist from Göttingen, uses this as a starting point for his theory on the origin of life.* In order to reach a certain length, polynucleotides have to replicate with high fidelity. This accuracy can only be obtained with the help of suitable catalysts: we may view them as copying machines. But these machines have to be built first. For this, a blueprint is required, which means information of some length: a few hundred letters seem to be the minimum. But such a length cannot be reached without the help of a copying machine. And this leads to *Catch 22*: no higher accuracy without a longer word, no longer word without a higher accuracy. It is the old scholastic riddle of the hen and the egg, after a face-lift from molecular genetics.

How did nature find a way out of this *impasse*? For the time being, one can only speculate. Apparently several short words must have combined to form the instruction for the first copying machine. A simple concatenation of these words cannot have been the solution, however, since it would yield nothing but a word too long to be correctly copied. Manfred Eigen and Peter Schuster proposed a *hypercycle* as a way out, that is, a catalytic feed-back loop whereby each word assists in the replication of the next one, in a regulatory cycle closing on itself. An essential prerequisite, the ability of polynucleotides to help propagate each other, has in the meantime been experimentally verified—undoubtedly a brilliant success. But there remain plenty of unsolved questions, and sceptical on-lookers may be forgiven for forming the impression that a solution to the problem of the origin of life seems to move ever farther and farther away.* Every attempt at an answer gives rise to more riddles. It reminds one of that notorious Greek monster which, whenever you slashed off one of its heads, grew two new heads in its place—also a kind of a chain reaction, incidentally. Which leads us back to branching processes. Let us take a leisurely look at their abstract backcloth—the theory of probability.

Odds and oddities

With the benefit of hindsight, it must seem astonishing that the first statisticians who studied branching could ever have stumbled upon the fallacy that extinction is *always* inescapable. If the average number of children is larger than one, then the average number of grandchildren is increased by this same factor, and so on. Thus the average number of offspring grows from generation to generation. Admittedly, the probability that the line has reached extinction also increases relentlessly. But can it grow to 100 per cent? If a population vanishes with a probability of 100 per cent, then its average size would also have to vanish—isn't this obvious?

Obvious it may be, but upsettingly, it is wrong. As a matter of fact, probability theory teems with paradoxical results, and quickly undermines all confidence in so-called common sense.* The theorems belonging to the mathematics of chance are not a bit less reliable than those of algebra or differential geometry, to be sure, but they can be extremely odd; this in spite of the fact that we feed our intuition from childhood onward on a rich diet of games of chance.

No civilization seems to have neglected the pleasure of toying with chance. The ancient Greeks, who liked to ascribe all good things to themselves, held that the throwing of dice was invented by the warriors besieging Troy. But in fact, the Egyptians had anticipated them: dice have been found in tombs from the first dynasty. Coins, however, came up first in Greece. Can it have taken long before they were used for Heads and Tails? Card games date back to the Middle Ages: and Gutenberg, while duly publishing his Bible first, made haste to print a set of Tarot cards in the very same year. The first lottery was held in the Florence of the Medicis. The age of enlightenment provided *roulette* (the invention of a French police officer); industrialization brought slot machines; and today, random number generators are used in many computer games. Hence there is no shortage of opportunities to get acquainted with Chance, and not just in a *haphazard* way; nevertheless, there are always surprises in store (see Box).

Since the beginning of modern times, scholars have computed probabilities, sometimes with highly improbable results. It started with Cardano, who was a gambler as much as a mathematician, and went on with Pascal, who led a less frivolous life, but couched a theological argument in the form of a bet. The range of solid applications increased only gradually: the insurance business, the theory of heat, biometry. . . . But nowadays, the foundations of pure and applied science rest squarely on probability theory. Major parts of physics, biology, and economics have fallen under its sway. Nevertheless, it is still accompanied by paradoxes and fallacies without number. They remind one of jugglery and deceit—but these were always to be found in gamblers' company.

The gentle art of drifting

Do you care for a small appetizer? Then let us go for a drift. The simplest way of drifting is the *random walk*.* It is usually explained by means of a drunk groping along a wall. With equal probability, he takes a step to the left or to the right (this 'he' is the only male pronoun my editor didn't object to). If he, then, keeps at it long enough, he is bound to return to his point of departure. But sooner or later, he will also reach any other point of the wall. He is not

A small sampler of paradoxes

1. The numbers 1 to 18 are inscribed upon the eighteen faces of three dice as shown. Although the total amount of points is the same on every dice (namely 57), the red die is better than the blue: the odds are 54 per cent that if the two dice are thrown, the red will show a higher value than the blue. Similarly, the blue die is better than the green die. And now the paradox: the green die is better than the red! Whichever die I choose, my adversary can do better. I am well advised to leave her the first choice.

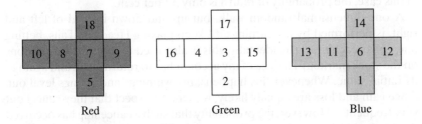

Red   Green   Blue

2. With two plain dice, one can obtain the sum 9 in two ways: as 3 + 6 or as 4 + 5. Similarly, one can obtain the sum 10 in two ways, as 4 + 6 or as 5 + 5. Why then does the sum 9 occur more often than 10? (One throws 9 in 11.1 per cent of all attempts, but 10 only in 8.3 per cent) By throwing three dice, one can obtain the sums 9 and 10 in six different ways. But this time, 10 is a more likely outcome than 9.

3. By repeatedly tossing a coin and waiting long enough, one can obtain any sequence, for instance Heads-Tails-Heads-Tails or Heads-Tails-Heads-Heads. In fact, the former sequence occurs on average once in 20 tosses, and the latter once in 18 tosses. So the former sequence is less likely to show up first, right?— Wrong. Heads-Tails-Heads-Tails is almost twice as likely as Heads-Tails-Heads-Heads to come up first.

4. It may happen that in a group of 365 people, no two persons have the same birthdate. With 366 people, this is no longer possible (if we exclude leap years): in this case, the probability of finding two persons with the same birthdate is 100 per cent. How large has the group to be to contain with a probability of at least 99 per cent two people with the same date of birth?—Most guesses are far out. The group has to contain only 55 members.

5. Two types of pills are compared. Allegedly, they cure headaches. First, male test-persons are used. Type Alpha helps 90 out of a group of 240 men testing it; type Beta helps 20 patients out of 60 men testing it. Hence Alpha is better than Beta—for men. Next, the pills are tried on female test-persons. Alpha is tested on 60 women, and helps 30 of them; while Beta is tested on 240 women and helps in 110 cases. Hence Alpha is better than Beta—for women. But altogether, 300 persons have tested Alpha and Beta each, and Alpha helped in 120 cases, Beta in 130—so Beta is better, after all! What should a statistician take for the headaches caused by this result?

very fast, incidentally, since he keeps changing his mind about which direction to take. Even the most sluggish snail will ultimately get ahead, if only it keeps creeping along: the drunk will at first pass it frequently on his wanderings up and down, but after some time, he will never again catch up with it.

If the wall were for some reason to tumble down, our drunk would have to take his random walk in two dimensions: he would step with the same probability to the north, the south, the east, or the west. Again, he will certainly return to wherever he started from. Surprisingly, however, this is different for a random walk in *three* dimensions (a dangerous exercise for the inebriated): in this case, the probability of return is only 37 per cent.

A one-dimensional random walk, but up and down instead of left and right, is performed by my earnings if I keep playing Heads or Tails, betting one dollar each time. (Incidentally, this is also used to describe fluctuations on the stock market.) The size of my account has to return again and again to its initial value. Whenever this happens, my winnings and losings level out. Since gain and loss are equally likely, we tend to expect that they cancel out very frequently. However, the probability that such a cancelling has occurred during the last half of the time I've been gambling is just 50 per cent; this is *always* so, no matter whether I've played for two hours or two years. Therefore it may take *a very long time* before winnings and losings level out once more. (In this game my account can take every positive or negative value; in reality, I will not get any further credit if my debts run too high; this means that my ruin is certain.)

Instead of playing Heads or Tails, we might also reach into an urn containing a red and a black marble. If we draw Red, we win, and if we draw Black, we lose one dollar. Before repeating the game, we have of course to put the marble back into the urn. Now if there were one red and *two* black marbles in the urn, the probability of a win would only be $\frac{1}{3}$. Again, the account will drift up and down, but with a strong overall tendency for going towards the bottom. This game is not fair. The winnings and losings may balance after a while, but this is no longer sure. What is sure is that after some time, our account will *never* surface again: the plunge is irreversible. The average frequency of winnings tends, as time goes on, towards $\frac{1}{3}$ (the proportion of red marbles in the urn).

If people are not told the proportion of red and black marbles in the urn, but are asked to bet on one colour, they should opt for Red or Black with equal likelihood. If, on the other hand, they know that there are exactly as many red as black marbles in the urn, they should again show no preference for one colour or the other. But if people are asked to choose between the two urns *before* placing their bet—the one urn with unknown composition, and the other with 1 : 1 proportions—they will very frequently pick the latter. The theory of probability cannot account for this. With the incomplete information given, the odds of winning are just the same in both cases. This so-called

Ellsberg Paradox does not properly belong to statistics, but to psychology. It throws a striking light on our instinctive reactions to risk and uncertainty.*

A notoriously counter-intuitive type of drift is obtained if, starting out with an urn containing one red and one black marble, we replace after each draw the marble we just picked and also add an extra marble of the same colour. Thus if we drew Red, the probability for Red in the next draw will have increased. Success, then, breeds success: the drift regulates itself. Such a positive feedback plays an important role in business and biology.* What will its outcome be in the long run? Again, the average frequency of winnings converges to a certain number (by itself, not an obvious result). But which number? Will it be $\frac{1}{2}$? Will it be 0 or 1, each with the same likelihood?—Most people guess wrongly. The truth is that at the outset, all values between 0 and 100 per cent are equally likely: a proportion between 2 and 3 per cent is as probable as one between 49 and 50 per cent or between 77 and 78 per cent. The first steps of this random walk settle its fate; the initial fluctuations lock themselves in (Fig. 4.3). The random events of the past affect those in the future. What started as a fair contest beween Black and Red can end up as a very one-sided affair. In every single such experiment, a *law* seems to emerge: after a few thousand draws, we can very confidently predict that the chances for Red to turn up are such and such. But for every repetition of the experiment (starting anew with one red and one black marble in the urn), a *different*

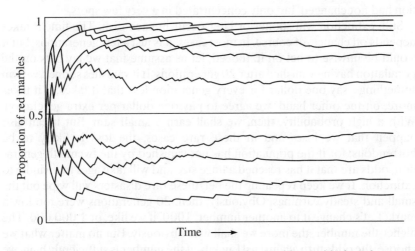

Fig. 4.3 Ten runs with the Polya urn. One starts out with one black and one red marble. After drawing a marble, one replaces it and adds a marble of the same colour. This is repeated 1000 times. Each path describes the varying proportions of red marbles in the urn. For each run, it converges on another value. The initial fluctuations are *locked in*.

law emerges: the such and such isn't the same. It is a recipe for how to obtain *irreproducible results*.

Dishing it out

Branching processes have their full share of paradoxical results. Let us take a look at the blackest sheep in their family, the *critical* branching process. What happens if the average number of offspring is *precisely* one (as, for instance, with bacteria having the same chances to duplicate or die)? In this case, the mean population number does not change from one generation to another. It remains just as large as the initial population was. But each such population will be wiped out *for sure*! The expected time till extinction, however, is ... infinite!

In order to understand this better, let us imagine lots of Petri dishes, with one bacterium in every dish. After one generation, about half the Petri dishes are empty, but the other half contains two bacteria. After 20 generations, almost all Petri dishes are empty—but in a few dishes bacteria have multiplied to such an extent that on average, we still have as many bacteria as there are dishes. Even those (by now densely populated) dishes are doomed in the long run; eventually, the last bacterium will pass away. But if we had started out with a thousand times more Petri dishes, there would still be some left whose population has not only survived, but multiplied to the extent of compensating for the disappearance of the others. It looks almost as if the total population had not changed, but only concentrated in a very few spots.

Still, extinction will sooner or later occur in each dish. The time it takes depends on chance, of course. In most cases, it is just a few generations; but it would be unwise to bet on it. Indeed, let us assume that we lay bets on the population having vanished after 20 generations. If it vanishes sooner, we win something—say one dollar for every generation less that it takes; if it takes more, on the other hand, we agree to pay one dollar per extra generation. With a high probability, then, we shall earn a small sum. But it can also happen that we lose—and in these rare cases, the loss is likely to be horrendous: for if the population has not vanished by the twentieth generation, odds are that it has reached a huge size and will take ages to dwindle to extinction. If we keep repeating the bet, these rare disasters will wipe out the small and steady earnings. Obviously, then, 20 generations were too low a mark. Let's change it to another number: 1000, if we like, or 1 000 000. The higher the number, the more we shall win, obviously. But no matter what we choose, the odds turn against us! Luckily, if the number is sufficiently high, we can be *dead sure* to be gone long before losing our first bet—but the debt falling upon our heirs would most certainly crush them.

Chance vs fortune

Let's halt before all credibility has gone. The laws of probability are not just a bunch of oddities. They have rightly been called *the true logic of the world* (James Clerk Maxwell), *the most significant object of our knowledge* (Pierre Simon de Laplace), *the most important concept of modern science* (Bertrand Russell), and other flattering expressions of the sort.* And Chance has also had its share of fulsome praise, in spite of being so awfully hard to define. Schrödinger said that chance is *the common root of the strict laws of nature** and Monod viewed it as *the foundation of the wonderful edifice of evolution.** But nevertheless, we seem not to have developed a firm feeling for the way it works, in spite of being exposed to its vagaries all the time. On the one hand, we expect from chance that it creates random fluctuations, and on the other hand that it averages them out. This relation between mean and dispersion is not at all easy to grasp. It is almost impossible, for instance, to *think up* a random sequence of Heads and Tails. Statistical tests will usually show that the series is far too well balanced to have been produced by *really* throwing a coin.

In the play of evolution, which has been running for billions of years, chance is cast for two parts. One type of chance is that of physicists, the other that of insurance brokers. One type keeps the molecules endlessly dancing, the other has bricks falling down from the roof. One works in the sub-microscopic domain while the other is as large as life. The former is what Schrödinger has in mind: the root of the physical laws is the superposition of countless random events. Such small-scale events are the causes of copying errors affecting the genome. They tie chemical bonds with the wrong nucleotide, or maybe break a chromosome in two . This yields mutations and recombinations, and hence a steady *supply* of genetic diversity. The other type of chance works in a cruder way: it causes accidents, transmits diseases, or brings together partners for life (it need not always be a mishap). This type of chance is often known as Fortune. It *restricts* the diversity of all possible designs to an arbitrarily selected sample. Both types of chance together form the *foundation of the wonderful edifice of evolution* which Monod praises so rhapsodically: *pure chance, nothing but chance, absolute blind freedom* . . .

The title of Monod's famous book is *Chance and necessity*, and therefore necessity must have its place somewhere. Now since mutations are undisputably random events, it is selection that is usually assigned to the realm of necessity. This is indeed the way *artificial* selection works. The conscientious breeder has a plan and carries it out with order and method. But *natural* selection has no aim and works capriciously. The catch-phrase 'survival of the fittest' creates an altogether exaggerated picture of its lawfulness. It helps, no doubt, to be well adapted to your environment; but for survival, you need to be fortunate too. It happens not rarely that the better boxer is floored by a so-called *lucky punch*; it will be small consolation for him that bookies had him eight to one. But at least he can hope for a second chance. Out of the ring, life is not so fair. No rule book, no referee. The fittest fish stands no earthly chance if its lake dries out.

We tend sometimes to make species responsible for having met extinction;* this seems to be a reflection of Protestant ethics and our cult of success. The much-abused dinosaurs would not have been much better off with larger brains and slimmer bodies; certainly not if the cause of their doom was the impact of a meteorite, and the subsequent darkening of the sky by a layer of dust. One can hardly be expected to cope with catastrophes of this calibre. Such *Acts of God* are explicitly excluded in the small print of insurance policies. What one *can* insure against are accidents happening often enough to allow for reasonable bets: storms, shipwrecks, etc. Events of this size do not wipe out entire species (except if they hit Noah's Ark), but they are able to nip the most promising mutation in the bud, and so to affect evolution in a decisive way.

Lots of bets and bad lots

Even the most advantageous traits do not *guarantee* success (which, in biology, means successors); we can only expect it to work *on average*. If, to stick to a well-worn example, an Ice Age approaches and winters get gradually grimmer, then the survival chances are obviously better for a wild horse with a thick coat than for its less protected brethren. But Siberian temperatures are not the only danger: the horse can tumble over a cliff or die in an epidemic before having had an opportunity to reproduce. If it happened to be the first horse with genes for a thicker coat, these genes will have passed away with it. But if thousands of well-coated horses lived side by side with thousands of thread-bare companions, then the deaths by accident will balance in the two groups, while the former will freeze less frequently than the latter. In this case it is fairly certain that the thick-coated line will have more offspring. Insurance companies, or bookies, for that matter, place a large number of bets and quite rightly feel safe. Of course they miss the thrill of putting it all on one horse.

Thus chance plays a role both for mutation and for selection. The difference is that chance is *undirected* for mutations. This does not mean that all mutations are equally likely—a mutation from A to B will usually not have the same probability as one from A to C, or even one from B to A. What is meant is that the probability for a mutation to occur does not depend on whether this mutation is advantageous or not, whereas the probability for the mutation to spread does so depend. (A change in the environment can increase the advantage of a mutant; on rare occasions, it can also increase the mutation rate; but the two events have nothing to do with each other.)

The interplay of advantage and chance in the working of selection can be followed in a simple thought experiment which belongs to the classical *repertoire* of population genetics.* Let us begin with a population of, say, 100 individuals. In order to avoid difficulties which have nothing to do with business at hand, we will assume that each individual can propagate independently of all others by simply spawning copies of itself. Thus we step deftly around the complications of sex and keep to mere chain reactions. In contrast to branching processes, however, we will now prevent populations from exploding without check—nature prevents them too, after all. Thus we select 100 from the offspring and eliminate the remainder (in thought). It follows that after one generation, the population is as large as before. And to be scrupulously fair, we select *randomly*: by distributing *tickets* among the offspring, for instance, and drawing 100 lucky winners. Only these winners can propagate.

So far, only chance is operating. Drawing tickets corresponds to *random sampling*. Each generation is therefore a sample from the offspring of the previous generation. This yields a recursive procedure which can be repeated ad lib.

In reality, populations are *not* constant, of course: but their growth is never unlimited. Most populations could produce many more offspring than their environment can sustain. The 'struggle for life' is rooted in this population pressure. Darwin actually discovered his principle of natural selection while reading (*for amusement*, as he writes) the grim text of Malthus on over-crowding and undernourishment.* The limited number of winning tickets corresponds to this limitation of living space.

Fixing on driftwood

To fix ideas, let us start with 100 marbles, some red and some black. We add to each marble another marble of the same colour (this is replication). We then throw the 200 marbles thus obtained into an empty urn, and shake it well. Next, we draw 100 marbles at random from our urn (this is the sample) and dispose of the rest. And this we do over and over again, starting each new round with the 100 marbles from our last sample. The proportion of red balls will increase and decrease randomly: it drifts up and down. But sooner or later, it has to reach the value of 0 or 100, for to avoid these values forever is as impossible as to keep winning endlessly at roulette. Once 0 is reached, or 100, the marbles are all black or all red. Now drifting is over: all subsequent samples will have the same colour. The probability that Red reaches *fixation* is as high as the proportion of red marbles in the initial urn (Fig. 4.4).

Let us assume next that all 100 marbles have different colours. To each one we add some more of the same colour (its *copies*), throw the whole lot into an empty urn, and sample 100. It could happen that we obtain marbles of all

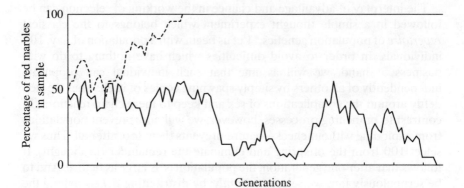

Fig. 4.4 Random drift. The 100 marbles from a sample are duplicated and thrown into an urn. One hundred marbles are drawn at random from the urn, to form the next sample, and so on. The proportion of red marbles drifts up and down until fixation is reached. Initially, the sample contains 50 red and 50 black marbles. Two runs are shown. (Redrawn from Maynard Smith 1989.)

colours again: but this is highly unlikely. Chance will hardly ever be so fair as to provide for each colour *exactly one* representative in the sample. After at most a few generations, one colour will have vanished from the urn, while another is present more than once, to make up for it. If we keep this going, there will be fewer and fewer colours in the consecutive samples. Finally, only one colour will remain, and this is when things freeze up. It will *always* happen that after a time, all members of the urn are descended from a single common ancestor.

Up to now, we have only watched *chance* at play: all colours were born equal. Now for *advantage* to enter the game. Such an advantage will be expressed by a higher rate of propagation. Let us assume for instance that every member of our population produces three copies, with the exception of one single mutant X giving birth to four copies. The privilege granted to X is tremendous. With unchecked growth, the mutant would, after a mere 16 generations, have more offspring than the 99 others together. But with selection through random sampling, it may happen that one generation contains no copy of the mutant at all. (It could be the very first generation: there will be 4 copies descending from X and 297 from all others, and odds are accordingly one in five that there will be no X among the winners of the first round.) Once this happens, X is eliminated, and future generations are not going to see it again. X is safe from extinction only if one of the samples, conversely, consists *entirely* of copies of the mutant. In this case X has reached *fixation*.

It is certain that one of these two alternatives will ultimately hold: X can reach fixation or extinction; there is no third way out. It is *all or nothing* for the mutant, and the verdict is without appeal. As soon as it has been reached, chance departs from the scene (until it manages to produce another mutant). But before the decision occurs, the number of X-copies *drifts* capriciously up and down. Mutation and selection thus offer a twin role for chance; some of it is captured by the two meanings of the verb *to err*. A mutation is a copying error; and its fate is determined by an erratic up and down. Evolution is the errings of errors, so to speak.

How high is the probability that X reaches fixation? In spite of its considerable advantage, it amounts to only 56 per cent. Its survival is not a safe bet by any means. But after all, there are 99 competitors to be elbowed out of the urn—hardly a pushover.

One might argue that the mutant could have found worse odds. A population of 100 is rather small. What about the expectations of the mutant in a population which is 1000 times larger? Would its survival chance not be 1000 times smaller?—Not at all. Surprisingly, it only changes by less than 1 per cent.* The probability that X reaches fixation is still close to 56 per cent, in spite of the huge increase in competitors all out to earn the urn (sorry!) for themselves. Which shows once more how difficult it is to guess odds off-hand.

Again, the outcome is settled after a few generations. If the offspring of a mutant have not been eliminated by then, it is most probably so numerous that the random swoop of the sample can hardly ever miss it any longer. Fixation is then practically guaranteed: whether the population is 100 or 100 000 000 does not affect the mutant's fate significantly.

But evolution, in the eyes of population geneticists, consists precisely of the occasional fixation of a new gene. Once a mutant has been eliminated, the game will proceed only after a newcomer steps in and starts a new bid for hegemony. It is similar to *Aggravation* (or *Ludo*), where you can enter a new pawn for the race only after throwing double sixes. Sometimes, it takes several rounds before anyone enters a pawn—fewer rounds, by the way, the more there are participants to the game. Similarly, in a population which is 1000 times larger, mutations occur at a thousandfold rate.

It follows that a change for the better (that is, the fixation of an advantageous gene) is more likely to occur in a *large* population:* indeed, the frequency of the occurrence of a beneficial mutation increases with population size, while the risk of its elimination is hardly affected by it. Again, this may seem counter-intuitive. Most people tend to agree that small groups can adapt more speedily. Here it is different: evolution in small populations occurs at a slower rate. Is Darwin on the side of the big battalions?

Incidentally, it is rather rare that a mutation carries an advantage of 4:3, as in our example. A big change is usually detrimental for the organism. A golfer will not use a big whacking stroke if she (or he) has reached the putting green: for even if the ball was hit in the right direction, it would sail across the top of the cup. And an organism is likely to be pretty close to the hole—or else what had adaptation been doing all this time? Selective advantages of the order of 1 per cent or 0.1 per cent may already be considered as bonanzas. For such small values, the probability of fixation is about twice as high as the selective advantage.* An advantage of 1 per cent has only a 2 per cent chance of spreading. On average, such a favourable mutation has to enter the population 50 times before it succeeds. The population has to wait a long time for it. But, as someone said to Marilyn Monroe in *Some Like it Hot*, it does not matter how long you have been waiting; what matters is whom you are waiting for.

Neutrals to the front

To view mutations as misprints can have its drawbacks. While it makes one feel—quite rightly—that mutations are far more likely to be harmful than advantageous, it yields the wrong kind of metaphor when it comes to mutations which are *neutral*, in the sense that they neither improve nor damage the text. Typographical errors can hardly ever be neutral, except possibly in nonsense poems or meaningless mumblings. Now the genome does indeed contain long paragraphs of so-called *junk* DNA which are not even translated. But neutral mutations can also occur in meaningful parts of the text. If, for instance, a misprint changes the RNA triplet GUU into GUC, this does not alter its function, which is to call for *valine* (one of the 20 amino acids used to build proteins). And even if the mutation creates a new protein, this can be just as good as the previous version.

How often neutral mutations can be observed in nature is a question to which we shall return. Let us stick a bit longer with the urn experiment, to see what happens to a neutral mutant. Again, its fate must be fixation or elimination—it cannot indefinitely keep up a tightrope act between these two options, in spite of the best-balanced neutrality. And at first, the career of a neutral mutant seems rather difficult to distinguish from that of a mutant with an advantage of 1 per cent. The probability that the offspring has been eliminated after seven generations is 79 per cent for the former and 78 per cent for the latter.* These odds can barely be told apart with the naked eye. And yet the end results are vastly different. The probability that the advantageous mutant reaches fixation (some 2 per cent, as we know) hardly depends on the size of the urn. In contrast, the fixation probability of the neutral mutant is just the inverse of the population size: in a population 1000 times larger, it is 1000 times smaller. It's elementary: if the mutant is neither better nor worse than the others, all individuals start with the same chance of having *their own* offspring reach fixation. There is equality of arms. For a population numbering 10 millions, the probability that the neutral mutation survives is only 0.00001 per cent, which is smaller than my chance of winning the jackpot in the national lottery by next Friday.

But hold it—that's not quite right! For I stopped trying my luck in a lottery (or anywhere else) years ago. Mutations, however, *never* give up. Since they keep on appearing relentlessly, they will sometimes succeed in replacing the previous version and establishing a new hegemony (Fig. 4.5). The probability that such a changing of the guard is completed in any given generation (the so-called *substitution rate*) is just as large as the probability that a neutral mutation occurs (the *mutation rate*).* This relation between an event affecting one individual (the appearance of a new misprint) and an event affecting the whole population (the final disappearance of the previous text version) is all the more remarkable as it is independent of population size: with twice the population, the frequency of neutral mutations is doubled, but their survival chance halved (and it takes twice the time for the lucky neutral to reach the haven of fixation).

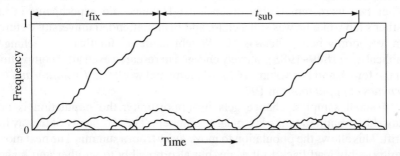

Fig. 4.5 Neutral mutants arise in a population. Their frequencies fluctuate. Most of them disappear, but some of them spread through the population and reach fixation (a frequency of 100 per cent). The average time t_{fix} needed by such an allele to reach fixation is proportional to the population size. The average time t_{sub} between two consecutive fixations (that is, the length of a period of hegemony) is the reciprocal of the probability that one individual mutates. (Redrawn from Kimura 1983.)

Shifty values

For deleterious mutants, survival is still more unlikely than for their neutral counterparts, of course: in large populations, their prospects are dim. This could leave us cold if they were not needed occasionally. Indeed, an improvement of the genome can sometimes be reached only through intermediate steps which are by themselves disadvantageous. This is like the Rubik cube, whose parts have to be twisted until every face shows one colour only. In order to get a better combination, one must destroy in part what has been achieved so far.

Evolution is frequently pictured as leading steadily uphill. If this were so, every summit in the landscape would be a *cul-de-sac*. Such a walk would end somewhere in the foothills of the mountain range. Whoever wants to reach the top of the range has to cross a few valleys on the way.

If I were to buy a lottery ticket, I would start by expending a small sum in the hope of cashing in soon. If I craftily offer a pawn to my chess partner, I hope to capture a rook in my second-next move. Such manoeuvres aim for an improvement in the long run at the price of a temporary disadvantage. But evolution has no aim. *Sheer chance* must be able to provide occasionally for the fixation of a deleterious mutant, in order to traverse a valley. From this point of view, a large population is seriously handicapped, for the apparently paradoxical reason that it offers no scope for mutants with handicaps. For such mutants, *small is beautiful*.

The American geneticist Sewall Wright was the first to spot the importance of small population numbers.* Together with R. A. Fisher and J. B. S. Haldane, Wright belongs to the holy trinity of population genetics. While the other two were remarkably precocious—Haldane did experiments in his father's lab while he was still a child, and Fisher attended university lectures on astronomy before he was 10—Wright made up for this by reaching a biblical age (1889–1988), actively engaged in research even after completion of the fourth and last volume of his monumental work* on *Evolution and the genetics of populations* in 1978.

In small samples, chance gets its opportunity: the population is not slavishly submitted to the law of large numbers. There is plenty of leeway for drift. This allows the population to move *away* from a summit. The next move can possibly lead back to it again, but also possibly to another and higher summit. The flexibility to take some steps downhill from time to time opens up a wide range. It is when the upward impulsion becomes a compulsion that evolution grinds to a halt.

In its initial stage every population is small. Later on, it may have to squeeze through other bottlenecks. It may also happen that small groups get split off, separated for instance by a geographical barrier from the remainder. In all such cases there are opportunities for the fixation of neutral or slightly deleterious genes. (Their disadvantage should not be too large, of course. If a gene reduces reproduction by as much as 5 per cent, its chances of invading a population with 100 members is practically zero.)

Which conditions are most propitious for the fast adaptation of a population? According to Wright, it should be of large size, but divided into many small subgroups, or *demes*, and offer a restricted interchange between these demes.* Such a structure makes use of the best of both worlds. The large size provides for the appearance of many mutations. If they are advantageous, so much to the better. If they are neutral or slightly deleterious, they will still have a sporting chance of fixing themselves in one of the small subpopulations; such a subgroup would then be able to 'cross a valley' and reach a higher peak. It would increase in size and could be used as a bridgehead for invading the other groups.

This *shifting balance* theory of Sewall Wright has met with a stern array of opponents. R. A. Fisher belonged to them. He proposed an alternative view, according to which the summits and troughs of the fitness landscape are submitted to endless ups and downs—he saw wave crests where Wright saw mountain crests. The endless stream of changes in environment keep the populations from ever reaching standstill. Large, well-mixed populations, many mutants with tiny advantages, and continuously changing conditions were Fisher's recipe.*

Reading the fine print

The *shifting balance* debate has been raging for half a century, and is far from being settled yet. But in the meantime, the main question is not so much which conditions would suit evolution best, but rather how it *really* proceeds. Since the 1960s, indeed, scientists have actually been able to decipher genetic information right down at the molecular level, whereas previously, they had to make do with its indirect expression at the level of the organism. The genome itself had been unreachable. Since all changes of the organs seemed to fall under the heading of *adaptation*, it was only natural, therefore, to assume that things would look no different for molecules.

The Japanese scientist Motoo Kimura shattered this easy belief. Kimura started his work in the United States, and was soon viewed as Wright's heir apparent. He later headed the National Institute of Genetics in Japan, in a magnificent location facing Mount Fuji. Kimura is very reserved, most at ease with breeding orchids, or tending his formulae, which have reached the sophistication found in theoretical physics. So mathematicians lean to his side in the great *neutrality debate* which Kimura's work sparked off in 1968. But meanwhile even geneticists from the opposite camp use Kimura's methods to dispute his claims.

According to these claims, mutations which are not immediately eliminated for their glaring defects are by and large neutral or slightly deleterious. Most fixations are caused by random drift and not by selective advantage. The effect of natural selection is in the first place negative: the prompt extermination of unfit mutants. The positive part of evolution—the stepwise restructuring of the genome in the entire population—is mostly the work of chance. In the overwhelming majority of cases, new mutations throwing their hat in the ring and trying their luck will be polished off. Occasionally, one will succeed, after a random walk lasting millions of generations. In general, it will not be any better than what it displaced. The march to fixation by a mutant with an advantage will occur much more rarely still (but proceed at a far faster pace).

It must be stressed that all this is intended to describe what happens at the *molecular* level.* It is not meant to contradict Darwin's views on adaptation by natural selection acting on organisms. We may compare this with the situation in physics. There is no contradiction between what we know about the levers and pulleys in the familiar world of elementary mechanics, and the weird small-scale structure of solid matter, with tiny atomic nuclei far apart from each other (if these nuclei were as large as golf balls, the distance between them would exceed the size of the largest golf course). Nevertheless, the fact that we don't fall through the floor requires some explanation.

The corresponding explanation for the fact that the random walks of genes can lead to marvellously subtle adaptations of organisms is not at all worked out—it is a long way from DNA chains to bodily traits; but most biologists feel (too complacently, as other biologists claim) that in essence the reason was known to Darwin already. It is based on the stepwise accumulation of small advantageous changes. The net result is tremendously complex: a single random event could never bring it about, just as a hurricane crossing a junk-yard could never assemble an aircraft (to use an oft-quoted slogan). But the chance of winning the lottery, while not high, is much better: we can watch it happen. Chance events of this sort, hoarded over millions of years, keep evolution going.*

Selection often rejects advantageous mutations, but this only slows down the pace. It accepts countless neutral mutations, but there is no harm in that either. Of course what is neutral today can turn out to be significant later. It is like altering a poker hand: the first card you draw may seem utterly worthless, but the card you draw next makes it an indispensable part of a straight. Similarly, a new genetic variant seemingly without immediate effect can reorient the whole future course: you may have to go for knaves, now, instead of kings.

Nevertheless, the neutral theory was felt to be a rival of natural selection. It aroused controversies which made the *shifting balance* argument look like child's play.

Still life with clocks

Kimura got started by the discovery that surprisingly many molecules exist in several variants within one population. Some individuals carry one variant, some another. Apparently such a coexistence of several variants marks a transitory stage which will last until one of the variants reaches fixation. From the frequency of these transitory stages, one can conclude that they last very long; for it they did not, we could observe them only rarely. Now if one variant were better than the other, the transitory stage would be fairly short, and thus easy to miss. But since it seems to take, on the contrary, a long time before a decision emerges, this speaks in favour of the neutral theory.

The final word has not yet been said. Critics of the neutral theory* stress that selection can keep several variants in the population if each variant becomes advantageous whenever it is rare. It is obvious that such a mechanism yields some protection against the danger of extinction. This type of selection plays an important role, and we shall meet it frequently in later chapters. But whether it is at work also at the molecular level has not been proved yet.

In order to check whether a given molecular type occurs in one or more variants, one has to analyse a representative sample of the population. This corresponds to a snapshot of the population at a given time. Snapshots of this sort, repeated every 10 000 years for hundreds of millions of years, would provide a film of the molecules' history: long periods of hegemony by one variant, interspersed with transitory phases which may lead to the substitution of one variant by another. During the same time, the organisms will also evolve: fins may turn into legs, scales into feathers, etc. Even if the filmmakers were exclusively interested in the molecule, they could not afford to ignore the organism's history; from time to time, a new species splits off, and the filmmakers must decide which one to keep tracking. In distinct populations, the molecule lives through different sets of adventures. The entrance of new mutants on the stage, their fortune, and their end, will proceed by different scripts.

Such films are not available, but we can compare snapshots of today's populations and infer something about the stories. The further back two species split off from a common ancestor, the less the present versions of their molecules have in common. The best-known example is haemoglobin, which consists (for humans) of a chain of 141 letters (or amino acids). The corresponding molecule occurring in the horse differs by 18 letters; in the chicken, by 35; in the carp, by 65, etc. (Fig. 4.6). This allows us to trace molecular descent and to construct evolutionary trees. They correspond reasonably well with those established by palaeontologists.

The truly surprising finding, however, was that the change in the haemoglobin molecule occurs at a fairly regular pace: like the ticking of a *molecular clock*. Each tick is the substitution of one variant by another. Some variants hold on for longer than others, of course, but the average length of their hegemony is the same, no matter whether it is measured for lines of fish, fowl, or game. Comparing the evolutionary rates of bodily traits, like the length of a bone, for example, leads to a strikingly different result: these contrast very much from one line to the other. Evolutionary rates of molecules do not.

For the neutral theory, this is easy to explain. We know that the rate of substitution of *neutral* mutants is as large as the rate of mutation, and this latter

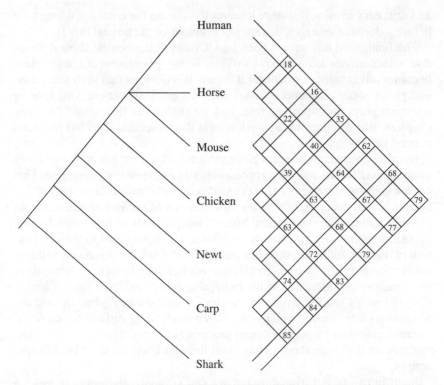

Fig. 4.6 Molecular clock. The number of sites with different letters (amino acids) for haemoglobin (more precisely, its alpha-chain) is shown for different vertebrates. The three mammals (human, horse, and mouse) differ by about 20 letters from each other. The shark differs by about 80 from all mammals. On the left is the evolutionary tree. (Redrawn from Kimura 1983.)

should be roughly the same for different variants of the haemoglobin molecule. For *advantageous* mutants, on the other hand, the probability of their fixation is given by the product of the probability that they occur (that is, mutation rate × population size) and the probability that they reach fixation (which is twice their advantage). It takes strong faith to believe that in lines as different as those leading to humans, chickens, or carp, the product of population size and advantage remains the same over hundreds of millions of years.

Playing fast and loose

The evolutionary tree was, of course, studied not just for haemoglobin, but for other molecules too. The results agree fairly well. But the molecules differ greatly in the *rate* at which their clocks are ticking. Cytochrome C, for instance, needs on average three times longer than haemoglobin for the substitution of a new variant; with fibrinopeptides, on the other hand, things happen eight times faster.

For the neutral theory, a beguilingly simple explanation is at hand. Not all parts of a molecule are equally constrained by its task. It is like with a key: that part which has to fit into the lock is fully prescribed, and each alteration makes it worthless. The other part is also necessary, for the key has to be handled and turned, but it is less strictly specified; within reasonable bounds, it can stand a lot of change. Similarly, parts of a molecule may be altered and others may not. If the task which a molecule has to perform does not allow it much freedom, it will evolve only slowly: for then, the probability that a change damages vital parts is high. If the constraints are not so tight, however, then changes will be tolerated much more easily.

With fibrinopeptides, for instance, which are involved in the clotting of blood, the precise structure seems hardly to matter: the molecule has plenty of leeway. It can mutate almost without restraint. The same holds for the so-called *silent mutations* in the DNA molecules of the genome. These do not affect the genetic instructions, either because they are not translated into the 'machine language' of the cell, or because the translation yields the same text as before. The evolutionary rates of these molecular parts are the highest we know. We find them whenever *no* selection pressure operates. It cannot be, therefore, that selection pressure is the cause of most fixations. The fast evolution of a molecule is due to the absence of constraints, rather than the presence of advantages.

The word *play* is sometimes used to mean the limited mobility of a mechanical part (like the French *jeu* and the German *Spiel* also). In this sense we may speak of a molecule's 'play' if some of its positions are less rigorously specified than others. Mutations can jiggle around freely with such loose parts. The looser the constraint, the faster the change. Conversely, one cannot afford to experiment under conditions of stress or dire need. Proverbially, one doesn't change horses in mid-stream; not even if a stronger team were at hand, ready to take over.

The neutral theory claims that most molecular changes reaching fixation are irrelevant. This has been used to insinuate that the theory itself is rather irrelevant for evolution: the relevant aspect being, of course, adaptation. The ceaseless change of shift performed by neutral mutants would accordingly be mere *evolutionary noise*. Some have qualified it as *boring*.

It is true that watching a game of chance is usually boring: but only if one has no share in it. We are not mere bystanders of neutral evolution: we are its product. The neutral theory appears to yield the only plausible scenario for the emergence of genes performing new tasks. In the first act, an established gene is *duplicated*. This mutation is more or less neutral (some extra material has to be copied, but copying comes cheap). It offers a playground, however, for further neutral mutations working upon the superfluous copy. Like a tinkerer who throws nothing away, the genome keeps useless stuff—the so-called *junk* DNA—on which it can harmlessly potter around until something of possible use turns up.*

Necessity is said to be the mother of invention. But abundance is a better mother. Want and necessity tend to rather conservative solutions. We see it in our cultural history: much of its progress is due to people freed from the need to worry about basic necessities: English country squires, for instance, or Athenian citizens. (Most of my colleagues would content themselves with a reduced teaching load.) And to go back somewhat further in the history of humankind: when our ancestors adopted an upright posture, several millions of years ago, they thereby freed their arms. As soon as hands were no longer used for support (in the *knuckle walk* of primates), they could be used to much better effect, for carrying things and using tools. *Homo* became first upright, then tool-handling, then clever: it all started with the loss of a constraint.

All about Eve

Speaking of ancestors: in 1987, Eve hit the news in a big way. It was revealed that she had lived some 200 000 years ago in Africa. Eve is, of course, a common ancestor of us all.*

We all doubtless share a male ancestor too (Eve's husband, if she had only one; and her father in any case). So why didn't we hear about Adam? Because the molecule which was used for the dating is *mt*DNA—a polynucleotide which does not, like most DNA, lodge in the cells' nuclei, but in the mitochondria, the power units of the cell. This molecule finds no place in the sperm: *mt*DNA is transmitted in the eggs only, and hence follows the female line. It is not subjected to the vagaries of biparental inheritance and sexual recombination. For this reason, it is easy to backtrack a molecule of *mt*DNA: it was inherited from the mother, who got it from her mother, etc.

There are billions of women around today. We know (from the urn model, basically) that copies of the *mt*DNA molecule of *one of them* will eventually reach fixation. Similarly, all of today's *mt*DNA descends from one ancestor. Ideally, the copies should all look alike. They do not because, in the meantime, mutations have been at work. The molecule is a fairly large chunk with a lousy replicating mechanism, so it evolves rather fast. The copies have diverged. But this can be used to trace the family tree from today's variants all the way back to Eve's original.

Eve was not alone, of course. She may have had some five or ten thousand female contemporaries. Many of them must have copies of their genes still around. It is only their *mt*DNA which has been elbowed out by Eve's. Just as we have two grandmothers and four great-grandmothers, we have many, many great-great-great ... great-grandmothers from Eve's generation. If I wrote out all those *great-*, the chapter would almost double in size; unfortunately, I'm not paid by the line.

Having the female line transmit those 16 792 pairs of letters of *mt*DNA appears as nature's way of making amends for our cock-eyed convention that family names are carried by the male line. Which leads us back to where we started from—as any proper random walk must do. But there is a further twist to Eve's story. In 1992, it became clear that the statistical method of constructing the family tree had been applied in a somewhat over-sanguine way.* Many other family trees appear to be just as likely as the one that was eventually put forward. This casts no doubt on Eve's existence, but she may have lived in another place and at anothe time. In fact, the *mt*DNA molecule might simply be too small for the required amount of detective work. In much the same way, if a coded message is too short, the smartest deciphering will not help.

What one needs is more data. Far and wide the most promising candidate seems to be the Y chromosome, which is carried only by males. Since it travels from father to son, we have uniparental transmission again, and can try to construct another family tree, this time leading back to Adam, of course, the man who, by definition, sired all present Y chromosomes.

Both Eve and Adam may give us some clue to human origins; but the odds that the two ever had a date together are infinitesimal. There was no Garden of Eden for them to meet in. Considering this, they both did pretty well.

Population genetics

5

5

Population genetics
The double game of chromosomes

> *I am the family face; flesh perishes, I live on.*
> Thomas Hardy, *Heredity.*

Armchair genetics

We have two parents, four grandparents, and so on. If we keep going, we find that we must have had, some 30 generations ago, more than a billion fore-bears. This cannot be right, since at the time, deep in the Middle Ages, the total human population was considerably less. We seem to have lost many ancestors along our way. The obvious reason for this is *inbreeding*. Someone who issues, like Wagner's hero Siegfried, from a mating between brother and sister has only one pair of grandparents, for instance. Matings between brothers and sisters are rare, but the pairing of more distant relatives, like cousins of the third degree, also leads to a loss in ancestry.

In some way or other, we are all interrelated, of course. But what do we have in common with whom? Family resemblances are a favourite topic of conversation. Children are relentlessly subjected to inspection by well-meaning relatives. I used to hate it, but that, according to my grandmother, was a family trait too.

In spite of all this widespread interest, it took a surprisingly long time to discover the laws of heredity. A hundred years before Mendel, all the elements needed for a solution were known. One can read about them, for instance, in the delightful *Vénus physique* by Maupertuis, the French president of the Berlin Academy of Science, a brilliant mind who earned a place of honour in the history of mathematics for his variational principle, and a footnote in the history of literature for his bitter feud with Voltaire. Maupertuis retraced in detail how uncommon traits, like albinism, were passed on. There was the case of one Elisabeth Horstmann from Rostock, for instance, who had six fingers on both hands. Her daughter inherited this painful handicap. Four of her eight children had six fingers, among them one Jacob Ruhe who became a well-known surgeon in Berlin. Two of his six children inherited the trait.

Maupertuis anticipated Mendel in at least three respects.* He considered traits—like the six-digitism—which were either present or absent, rather than continuously graded like the shape of the nose. He followed their transmission over *several* generations. And he made use of statistics. He certainly came very close to a solution—in retrospect, one gets the feeling that the great man would have only needed to lean back in his armchair.

It is quite possible that Gregor Mendel made his great discovery in an armchair. He left no notes on this topic, but R. A. Fisher, who scrutinized Mendel's papers, wrote that many of the decisions made by Mendel in planning his experiments must seem incomprehensible if one does not assume that he knew full well what to expect; that those experiments were meant as confirmation of a theory obtained through an entirely abstract approach. Fisher even sketched this approach.* Let us see how it would have worked out for Maupertuis.

The case of the six-fingered surgeon

In the first place, a case-study like that of Jacob Ruhe's family shows that a child can inherit traits just as well from its father as from its mother, and this irrespective of its own sex. So much for the *spermists* and *ovists*, who favoured one sex over the other. Now since both parents are involved in transmitting traits, it is simplest to assume that they do so to an equal degree; which leads to the hypothesis that a child's trait comes with the same likelihood from the father or from the mother.

This, however, is wrong: in fact, it is the major pitfall of the reasoning. Everything about the six-fingered family could be explained in this way, but the transmission of other traits shows that it cannot cover the truth. It was well known to Maupertuis, and in fact already much earlier, that some traits can vanish, only to reappear after a few generations. A blue-eyed child can have two brown-eyed parents, for instance: it inherits the blue eyes from more distant ancestors. In such a case, some hereditary factor for blue eyes must have been transmitted by the parents without having been expressed in their bodies; some other factors have been expressed, causing them to have brown eyes.

Hence offspring receive more than one factor per trait. It is simplest, again, to assume that they receive *two* factors, one from each parent. Now it follows by elementary bookkeeping that whenever a child, in turn, produces an offspring, it can pass on only half of what it has received; it transmits only one of the two factors. Apparently, this factor comes with the same likelihood from the father or from the mother. Indeed, in one generation of the Ruhe family, four out of eight children inherited the six digits from the mother, while in another generation, two out of six inherited them from the father. This is much as if a toss-up had decided.

To repeat: the simplest hypothesis is that each trait is caused by one factor.

But the re-emergence of traits after several generations excludes this. The next simplest hypothesis is that each individual inherits one factor from each parent, and transmits one of its two factors to each offspring. This leaves two open questions:

1. Which of the two factors is passed on? Observation suggests that both have the same chance.
2. Which of the two factors is expressed? Observation suggests that some factors are *dominant* and always overrule the others.

It could have turned out to be the other way round: that chance decides which factor gets expressed, and that some factors always take precedence in being transmitted to offspring. But information already available at Maupertuis' time gave hints to the right answers. The proportions of six-fingered children bespeak of equal chances. The fact that brown-eyed parents can have blue-eyed children suggests that the factor Brown is dominant, whereas Ruhe's family story suggests that Six is dominant.

The next task must surely be to test whether these simplest answers hold. For instance, three out of four children born to two six-fingered parents should be six-fingered. But this, of course, is rather difficult to test. Mendel devised more feasible experiments* (Fig. 5.1), drawing on statistical methods and on techniques of crossing and breeding which had not been available to previous centuries.

Incidentally, both statisticians and breeders had traditionally been interested, for obvious reasons, in continuously varying traits like body weight or milk yield; but all knew that inheritance is blurred for such traits. Mendel cleverly used for his experiments only traits which occurred in well-distinguished characters in his peas, similar to eye-colour or number of fingers in humans. These traits were:

1. seed colour: yellow or green;
2. seed shape: round or angular;
3. the colour of the seed-coat: white or brownish;
4. the colour of the unripe pod: green or yellow;
5. the shape of the ripe pod: smooth or wrinkled;
6. the position of the flowers: axillary or terminal;
7. the size of the stem: large or small.

Even this last trait does not offer a continuous spectrum of variations: the stems of one class of peas were shorter than 60 cm, those of the other longer than 180 cm. Mendel explicitly allowed only for characters allowing a 'sharp and certain separation'.*

In all his experiments, which extended over six or seven years, the different crossings and recrossings yielded the right kind of frequencies—almost embarrassingly right, as statisticians later pointed out: usually, numbers don't

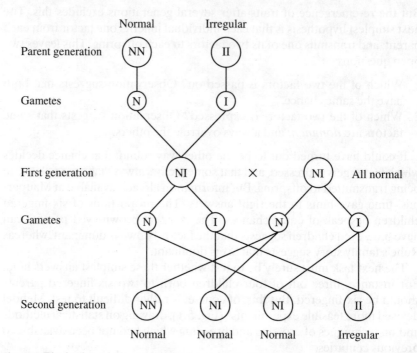

Fig. 5.1 The basic experiment in genetics, repeated here not for Mendel's peas but for Darwin's snapdragons. The normal type with the two alleles N is crossed with the irregular type having the two alleles I. All offspring have one allele N and one allele I and look like the normal type, since N dominates I. But when two plants of this first generation are crossed, chances are 25 per cent that their offspring inherits two alleles of type I and is irregular. In this second generation, the odds for normal plants are 3:1.

come out so pat. This uncanny fit* need not be ascribed to deception, however, or an Augustinian version of 'the end justifies the means'; it could be that Mendel, knowing the right answers all along, stopped counting his thousands and thousands of peas when the values satisfied him particularly well—a methodological peccadillo for which his contemporaries would hardly have blamed him. As it was, they didn't take any notice at all.

Blendingly obvious

For a long time, hereditary traits had been supposed to be carried in the blood. 'The near in blood,' to quote Shakespeare, 'the nearer bloody'. Expressions like 'blood will tell', 'kindred blood', 'blue blood', or 'tainted blood' were much more than just figures of speech. They suggested that inherited characters could mix in arbitrary proportions and vary continuously, like the colour in a cup of coffee.

Mixing, however, has the property that it is awfully hard to undo. Blending inheritance would be *irreversible*, in that case: but this is ruled out by the most elementary observations. The reappearance of characters after several generations is a clear hint that they do *not* fuse forever with the contribution from the other parents. They are just temporarily masked. But no one before Mendel understood this game of hide-and-seek.

Mendel had studied physics and chemistry and was well versed in what at the time was named the *atomic hypothesis*. His hereditary factors, which are called genes nowadays, correspond to atoms of inheritance: they are well-defined, durable units subject to statistical laws. In order to think of genes, one probably has to believe in atoms first.

Genes for eye-colour do not mix in the offspring. They separate in the next generation. This *segregation* was Mendel's main discovery. It would have saved his contemporary Charles Darwin from a most painful quandary. Natural selection needs to be provided with genetic variability to work upon—or else, what to select? But if the parental characters blend, variability decreases rapidly. It is halved from one generation to the next, if the mating pairs assort randomly. If they prefer to assort like with like, this decrease is somewhat smaller, but quite effective still. All diversity in a population would cancel irreversibly after a few generations.*

This can be seen in a thought experiment, by following the fate of a new mutant sporting green hair, say. All its possible mates would have hair of a more conventional colour. If characters fused, the mutant's children would have half of the green factor, the grandchildren one quarter of it, and so on; the new quality would get lost in ever widening circles of paler and paler copies.

With Mendelian heredity, things work differently. Here, the gene will *not* become diluted. Descendants can inherit it or not, with equal probability. A fine distinction: offspring do not get half of the gene; no, it is half of the offspring who get the gene (statistically half, to be precise).

The odds that the gene will spread increase if the number of offspring is above average. Maybe the green hair colour bolsters reproductive success; maybe not. But in any case, natural selection will have something to act upon. With blending inheritance, this would not be the case. Even if the most fleeting shade of green exerted an irresistible spell on the other sex, it would not help. The green would dissolve like a drop in the ocean.

Darwin came tantalizingly close to solving the problem of blending inheritance.* In fact, he almost duplicated Mendel's most basic experiment in his own studies on the flower shape of the snapdragon (Fig. 5.1). He had crossed a line with normal flowers and a line with irregular flowers, and found that all offspring had normal flowers. Then, in crossing the offspring, he obtained in the next generation 88 normal and 37 irregular plants, which is reasonably close to the 3:1 ratio.*

'Let theory guide your observations' was one of Darwin's maxims.* In his experiments with snapdragon, he observed the same facts as Mendel, but he did not have the same theory. What was the reason for it? As a contemporary of Maxwell, Darwin was surely well acquainted with atomistic theory, and he must have known about statistics from his cousin Francis Galton, who was foremost in this field. Mendel's only advantage, as far as one can see, was his training as a mathematician. Darwin often regretted, in fact, that he had not proceeded far enough to know 'something of the great leading principles of mathematics', and wrote that persons 'thus endowed seem to have an extra sense.'*

To return to the general problem: since genetic variants cannot persist under blending inheritance, Darwin felt obliged to conclude that most of what we see of them today must be of very recent origin. But how can such a large-scale supply of novel characters be kept up? For want of a better reason, Darwin charged it on the account of ceaseless environmental changes producing endlessly new adaptations. Hence Darwin allowed for the inheritance of acquired characters. Today, this view is branded as a gross violation of Darwinism, a fact which Darwin could not have foreseen. Still, he clearly felt that he was on treacherous ground; he just saw no other way out.

Nowadays all biology students are told that Mendel's discovery was just the missing piece in Darwin's puzzle. Did Mendel ever get an inkling of this? He certainly was well aware of Darwin's work. His copy of the *Origin of species* is filled with annotations; and the minutes of the Naturforschender Verein in Brünn reveal that on the winter evening when Mendel read his paper to the Society, the learned members received it in silence, and then turned to discussing the *Origin*.* But nothing seems to indicate that Mendel was aware of the role of his laws for the theory of evolution.

The *synthetic* theory of Neo-Darwinism got its (somewhat cumbersome) name from the claim to provide a synthesis of the theories of Darwin and Mendel. And so it does. In a way, it has to, since we cannot dispense with either. And yet it started out as a shotgun marriage, and not just because the two main parties had never been asked for their consent. In fact, the re-discovery of Mendel's laws was initially seen as a fatal blow to Darwin's position. The two grand theories, one stressing hereditary conservation and the other evolutionary change, seemed hardly reconcilable at first.*

The simple art of double dealing

The principle of natural selection is so simple that not a few philosophers believe to this day that it is a tautology. A hereditary trait allowing an individual to produce more offspring will be represented more frequently in the next generation. 'How utterly simple', as Darwin's great supporter Thomas Huxley is supposed to have groaned. 'How extremely stupid not to have thought of that.'*

The average number of offspring—the so-called fitness—will increase in the population, because variants with less than average fitness drop out. This has been compared to a high jump competition: in every round, the bar is raised some more, and the field of competitors shrinks. The bar denotes average fitness. Only those who surpass it can spread. The best variant is bound to win. But in contrast to the usual competition rules, new participants can join the game at any time, provided they do measure up. These are the variants introduced by mutations.

This kind of parallel processing of a whole population yields a simple recipe for progress, which is not limited to living beings. Manfred Eigen and his co-workers, for example, have studied the evolution of RNA molecules* in a test tube, and verified experimentally that the interplay of mutation and selection leads regularly to those molecular variants which can be copied fastest under the prevailing conditions. There is no need—and, what counts more, no time—to evaluate all *possible* molecular combinations; their number is far too large. Selection quickly leads to excellent solutions, because the new variants are produced from those which have proven their worth by having survived so far.

In a similar vein, engineers can improve their designs by simulating natural selection. They start out with a large array of variants and evaluate them; the better a variant, the more 'offspring' it provides to the next artificial 'genera-tion'. Not all of these successor models are exact replicas: some of them are modified randomly. A few rounds of trial and error are enough to produce excellent profiles for jet valves and optimal curvatures of flow pipes.

In nature, this method of not-quite-exact replication has led pretty far—roughly speaking, from simple RNA molecules on the threshold of life to today's bacteria, which are doing quite well. But it has not led to flowers, trout, humans, bees, and birds. Indeed, it relies on the fact that (barring mutations) all offspring consist of identical copies.* Some 1.5 billion years ago, however, a second, and very different method evolved, which trades simplicity for duplicity.* Higher organisms no longer replicate, in general. Apart from identical twins, clones, or vegetative sprouts, they are all genetically unique. Nothing ever repeats. Even the most successful blueprint is promptly torn to pieces—a bizarre way of handling (and eventually improving) a design.

The Mendelian mechanism subverts replication. Most higher organisms use two sets of instructions—one from each parent. No longer do they produce their like. What they produce are gametes—utterly different beings carrying single sets of instructions. These gametes do not replicate either. On the contrary, they fuse pairwise to form beings with double sets of instructions again.

Genes travel, in turns, in carriers which are *diploid* (with double sets) and *haploid* (single sets). We tend to view the former as the principal stage. But just as we use gametes to produce our offspring, so the gametes use us to produce more gametes. In the chain of generations, they are no less important than we are. Each gamete is a unique being, differing from all others by its immunological properties, its degree of mobility, its stock of provisions. No doubt, in our type of life-cycle the diploid form is more elaborate. With mushrooms, however, it's the other way round; and drones, which are haploid, seem not to be much worse off than their diploid sisters, the working bees (except, of course, for their proverbially low job consciousness).

Much as we have two parents each, so have our gametes. The genes which they transport come in part from an egg and in part from a sperm, which fused when we were conceived. These cells—egg and sperm—can be viewed as the gametes' mother and father. While we diploids receive our genes in pairs, one from each parent, the gametes receive only single genes, which stem with equal probability from their 'mother' or their 'father'. Hence gametes use the simpler of the two possible inheritance mechanisms described in our little exercise in armchair genetics, whereas we use the other one.

Double bookkeeping

Let us imagine that organisms try to tell their offspring all they will need for life in the form of a letter of instructions: *How to become human* or *A little guide for the pea.* Each such correspondence course must always address the same questions, which have been catalogued and numbered by an obliging organization. The parents have to fill in stacks of questionnaires for their offspring, like passport application forms. Size of the stem: *small* (one of the two officially admissible answers); seed colour: *green*, etc. The number of the question denotes the gene locus (or position on the genome). The corresponding answer is the gene. For each question, only a restricted set of answers is possible—the so-called alleles at this locus.

Diploid beings receive two letters of instructions, one from each parent. Parents are not always of the same opinion. The mother has ticked *large* and the father *small.* Such differences occur at hundreds of loci—in humans for about 12 per cent of all questions.

The messages are so long that the mail will refuse to handle them. They have to be divided into several separate letters—the chromosomes. The correspondence course on *How to become human* consists of 23 letters, for instance.

Sooner than expected, one may have to send off instructions oneself.

Rather than write something up, one just copies the collected letters from mother and father, and arbitrarily selects one from each pair. This we can do in 2^{23} (or roughly 8 million) different ways. In fact, the diversity is even larger. The letters will not all be sent off as we received them. It can happen that while copying them, we exchange a stack of questionnaires from father with the corresponding stack from mother. This *cross-over* produces new letters by *recombining* the parental answers (Fig. 5.2). Letter XII in one of my sperm can have the first 411 questionnaires from letter XII by my father and the remaining 974 questionnaires from letter XII by my mother. In this way, two questionnaires which were in the same letter can go their separate ways. The probability of this happening gets larger the farther their numbers are apart. The different chromosomes are totally unlinked. Answers on different chromosomes are freely recombined. If my sperm cell received chromosome XII from my father, this makes it neither more nor less likely that chromosome XIII comes also from him.

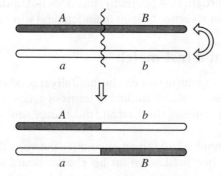

Fig. 5.2 A crossing-over. Two chromosomes exchange parts.

The number of chromosome pairs varies from one species to another. Peas, for instance, have seven of them. Since Mendel had studied seven traits, and found independent assortment, it seems as if he uncannily managed to choose one trait for each chromosome (the odds for which are less than 1 per cent). This looks like heavenly guidance for the future abbot, but as it turned out, the seven loci sit on four chromosomes only.* The loci are so far apart from each other, however, that they were routinely unlinked by cross-over. (Except, that is, for pod shape and stem height, a pair of characters for which Mendel did *not* check independent transmission. One can only speculate whether evidence for linkage would have troubled him.)

By the time Mendel's rules were rediscovered at the turn of the century, the 'minuet of chromosomes', their beautifully regulated matching up and splitting asunder, had been revealed under the microscope. It was almost inevitable that these little sticks would be suspected of being the carriers of heredity.* Segregation was now *seen* to happen: chromosome numbers were reduced to half in the gametes, and restored to full count at fertilization. Mendel had not needed a microscope; apparently, he was endowed with that 'extra sense' which Darwin ascribed to some mathematicians.

The possibility of shuffling chromosomes and parts of chromosomes creates an endless supply of genetic novelty. It does not alter the answers themselves—this happens only through errors in copying—but it combines them in ever changing ways. All genetic novelty stems ultimately from a mutation. But mutations are relatively rare. Recombination amplifies them. A mutation which produces a new answer yields only *one* new text among many; recombination, however, can *double* forthwith the number of possible texts.

Hence Mendel's inheritance scheme provides all the variability required for natural selection. But at the same stroke, it destroys even the most successful genetic program. Apparently, this does not undermine but rather boosts selection. But it seems difficult to understand why this is so.

Not cricket, but a test match

If it is no longer the organism that can be faithfully reproduced, but the single gene, then evolution should be studied in terms of genes. This is the vantage point of population genetics, that first and most successful alliance of biology with mathematics.

Its birth is commonly dated to an encounter between the biologist R. C. Punnett, who later became known for his checkerboard squares (Fig. 5.3), and G. H. Hardy the mathematician. It happened in 1908, during a cricket match.* Punnett wanted to disprove the claim that a newly produced dominant allele would spread until it was as frequent as its recessive counterpart. Hardy saw right away that allele frequencies will not change at all. The probability of inheriting a particular allele from a parent is exactly as high as the probability of passing it on to a child. If, for instance, one third of all genes in my parents' generation said a, then the probability that I inherit it from my mother is one-third, as is the probability that I receive it from my father. No matter if I pass to my son the gene which I got from my father or that which I got from my mother, the probability that it says a remains one-third. The child's probability of inheriting two a alleles is one in nine, of course. And this is it. In the spirit of cricket, the rule was called a law.

Since Hardy lived for complex integrals and trigonometric series (aside from cricket), he felt it as an irony of fate that his name had entered the Hall of Fame of genetics, forever tied to an argument somewhere between obvious and elementary.* On top of this, he had to share the glory with a little-known

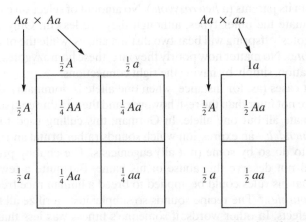

Fig. 5.3 Punnett squares (or checkerboard squares) showing that if a heterozygote having one allele *A* and one allele *a* is crossed (a) with another heterozygote or (b) with a homozygote having two alleles *a*, half of the offspring is heterozygous.

German physician named Weinberg, who had actually published the law six months earlier, but in the *Jahreshefte des Vereins für vaterländische Natur-kunde in Württemberg*,* where it was hidden almost as safely as Mendel's work in the *Verhandlungen des naturforschendenVereins in Brünn.*

Let us stress that this Hardy–Weinberg law presupposes random mating. If, for instance, blue-eyed people felt particularly attracted to blue eyes, matings would not really be at random. (At least not for eye-colour; for blood group genes, of course, it would still be at random.) If, to give another example, matings in the USA were independent of skin colour, then the white population would drop from 90 per cent to 48 per cent, while the population with darkest skin shade, currently half a million people, would practically disappear. Altogether, some 97 per cent of the population would have a very light skin; essentially, America would pale.*

For the Hardy–Weinberg law to hold, it must also be assumed that the genes in question do not affect survival and fertility. But what happens if they do? It is here, actually, that genetics becomes most interesting: in the inter-play between selection and heredity. The law of Hardy–Weinberg need not be precisely valid, in this case, although it may still offer a good rule of thumb.

Let us take a look at a simplified model, where individuals differ at one gene locus only. Depending on the alleles inherited at this locus, an individual may have more offspring, or fewer—a higher or a lower fitness. We know that in haploid organisms, the fittest type would prevail: after many a generation, all individuals would be of this type. For diploids, this does not always hold. It could happen that the fittest type is one which has inherited two *distinct* alleles from its parents (a *heterozygote*). No amount of selection can in such a case eliminate the other types, although they are less fit. Only half of the heterozygotes' offspring will bear two distinct alleles, while the others will be *homozygotes*. No matter how poorly they fare, these homozygotes will stay in the population, simply by having the right connections.

In most cases (as, for instance, when one allele is dominant), the hetero-zygotes are not fitter than the rest, however; and then, in due course, selection will eliminate all but one allele. In German, this culling used to be called *herausgemendelt*—an expression which sounds rather brutal and was, I fear, intended to do so by some (not all) eugenicists. The chubby prelate from Brünn did not deserve this abuse of his name. But quite a few scientists thought that his rules could be applied to breed a human race free from so-called *impurities*.* The recipe sounds so simple: just sterilize all those with genetic defects. In other words, if someone's fitness was less than 100 per cent, reduce it all the way to zero. A human sterilization programme is a horrible crime, of course, but quite in line with other Nazi measures. Some *Reichsamt für Erbhygiene* would see to it, and purge the Volk of all unwanted traits.

But let us compute. What would happen if a top-ranking eugenicist was struck by a violent aversion to albinism, for instance (which is not a 'defect', by the way), and decided to sterilize all albinos forthwith? One person in 20000 is an albino (as Maupertuis estimated already). But the allele for albinism is recessive. It works only in double dose. From Hardy–Weinberg it follows that one gene in 140 is such an allele, since 140×140 is (roughly) 20000. Since an offspring can receive this allele from each parent, one person in 70 will carry it. Sterilizing all albinos results in sterilizing only one in 280 bearers of the gene. To reduce the frequency of albinism by one half—a rather modest aim for our eugenicist—a rigorous sterilization programme would have to be applied during some 60 generations,* which is considerably longer than the 1000 years planned for the Third Reich.

With haploids, this would be different. A sterilization programme would eliminate any undesired allele within one generation. With diploids, an allele can hide behind another, dominant allele. Selection loses a lot of its punch.

The road to dominance

What causes dominance, then? It depends. Sometimes one allele says some-thing effective—like *turn on production of enzyme XY*—and the other says nothing at all. In this case, the active allele prevails; as Sewall Wright put it, 'if either the dominant or recessive is inactive, it must be the latter.' Having two copies of the active allele acts as a *safety margin* (the expression is Haldane's), much as two eyes are better than one.*

In a similar vein, if one allele whispers (metaphorically speaking) and the other speaks with an overbearing voice, it is the latter which will be obeyed. But this explanation cannot cover all cases. What happens if both alleles speak loud and clear and both say something meaningful? Here is where R. A. Fisher stepped in. According to him, *another* locus may decide.* The question *what is one to do if one answer at locus 666 says A and the other says B?* is perhaps addressed on locus 690. And one possible answer—a so-called modifier allele—says, *in that case, read A.*

This modifier theory has been verified experimentally by Fisher and others. They succeeded in selecting for modifier alleles, by breeding from one common stock of mice one line where the modifier said *read A* and another line where it said *read B* (here *A* stands for 'long tail' and *B* stands for 'short'). Dominance modifiers are, in fact, rather common; this is firmly established by now.*

What is less clear is how these modifiers were selected under natural conditions. In Fisher's scenario, the best allele is firmly entrenched, having practically reached fixation, whereas less good alleles keep being produced by mutation. Since they are rare, the probability that an individual carries two copies of them is extremely small; most will occur together with the common allele. A modifier saying that in such a case the common (good) allele should prevail will fare better than other modifiers, and will spread. Its advantage, however, is very small, since it works only in heterozygotes, which form a tiny minority. Fisher concluded, unflinchingly, that advantages of the order of one thousandth of 1 per cent, or even lower, can play a role in natural selection. But this is where many biologists part company with him.

In fact, other scenarios seem more plausible nowadays. The good allele must have been rare when it was first produced. In its early stage, when it was up and coming, a modifier making it prevail would have rendered it tremendous help, and been offered a lift in return. It would have assisted a winner in spreading, and not (as in Fisher's set-up) helped a loser in masking its defect.* More muscle to the good answer, rather than a muzzle to the bad.

Another road to dominance has been proposed by P. M. Sheppard.* Suppose that those individuals having two *distinct* answers are best adapted; this happens. As we have seen, such individuals will become very common, but cannot reach an absolute majority, since half of their offspring must have a double dose of the same allele, either *AA* or *BB*. A modifier causing *AB* to act like the less good *AA* will make a thorough nuisance of itself, and be shunted away. But a modifier helping an *AA* individual to act like the better *AB* will be highly welcome. In fact, it will have many opportunities to do good works, since *AB* is so frequent, and it will spread with great rapidity. The allele *A* will now always be good to have, whether in one or in two copies, while the allele *B* will be penalized whenever it occurs in a double dose. Hence, *A* will spread up to the point where almost all individuals will be of type *AA*. It seems that only this mechanism can account for the tremendous rate at which the dominance of 'black' in the peppered moth spread during the Industrial Revolution, when the bark of the trees in the Midlands grew darker.

Nothing recedes like progress

In spite of the Mendelian complications which allow recessive alleles to hide their true nature from sight, the average fitness of a population appears nevertheless to grow from any generation to the next. More than that: the larger the variance in the fitness, the faster its average grows.

In contrast to the haploid case, this is far from easy to prove for diploids. R. A. Fisher was rightly pleased with his result. He called it with becoming modesty the Fundamental Theorem of Natural Selection.* It appears to guarantee progress in evolution, and is not infrequently hailed as a counterpart to the second law of thermodynamics, which states that entropy must increase. So we see, in physics, disorder growing inexorably in systems isolated from their surroundings; and in biology, fitness increasing steadily in populations struggling for life. Ascent here and degradation there—almost too good to be true. 'My own view', as Haldane's heir John Maynard Smith wrote drily, 'is that [the Fundamental Theorem] cannot play an important role in biology.'* And in fact, its validity hinges on far too many unrealistic assumptions.

The first catch is that fitness does not just depend on the probability of reaching maturity; it also depends on fertility, which is not a property of the individual alone, but of the mating pair. If both these effects are taken into account, it can happen that fitness oscillates up and down: instead of progressing, such a population would cycle.*

Furthermore, the fitness of a trait may depend on its frequency. In many cases, it is good to be rare; in others, it helps to be common. An immunotype, for instance, loses efficiency when it is frequent: for then, microbes will adapt to it. On the other hand, a butterfly tasting horrible is better off if many members of its population carry the same distasteful trait, because then, birds will be more likely to know. But no matter if better when common or rare, if a character's advantage depends on its frequency in the population, then the mean fitness need not always grow.

The major restriction in deriving the Fundamental Theorem, however, is the assumption that fitness depends on *one* genetic locus only. In reality, it depends on many thousands of loci. Even if we model two loci only, we plunge into a thorny thicket of alternatives rarely leading to optimal fitness.* No steady improvement can be expected when several loci interact. Progress seems to have been sacrificed for the sake of diversity.

How seriously must one take these results? Many biologists shrug them off as petty quibblings. They *see* that natural selection copes wonderfully with the laws of heredity. If mathematicians are not able to come to grips with it, that's their problem. Oscillating allele frequencies have never been observed in laboratory or nature, so why worry? While models with two genetic loci are hard to deal with, they are not much closer to the full genome than models with a single locus only. If the whole gene complex were taken into account, things would work out. They would have to. After all, we are here to prove it.

Well, maybe. But for how long? It could easily happen, for instance, that our intelligence (which few would doubt to have been sharpened by selection) will allow us to exterminate ourselves. And we shall encounter in the following pages some selective processes which, far from improving a species, can undermine its survival.

Distorted characters

Just as a jumbo jet can fly although its parts do not, so the whole genome has properties which are not found in any of its components. Many biologists like to warn of reducing the glorious whole by merely collecting imperfect models of disconnected subsystems. The most striking aspect of an organism is, after all, that it integrates so many disparate functions in one coordinated unit.

But parts of the whole tend sometimes to go their own ways; such malfunctions frequently reveal more than the unruffled face of normality. Hidden tensions work in each genome, remnants of a long history of repression and constraint.

We know precious little about the momentous step which led from haploids to diploids some 1500 million years ago. It could have originated in the curious habit of *conjugation*: two bacteria may dock side by side and transfer parts of their genomes.* If this exchange is imperfect, one bacterium will end up with double sets of genes. If its descendants are reduced to having single sets again, these can be viewed as its gametes; and if, at a subsequent conjugation, one such descendant injects its genes into another, this would correspond to the fusion of two gametes to form a zygote. This is admittedly vague; nobody really knows. But at the beginning, the motions of chromosomes cannot have been as smoothly choreographed as they are now.

We get a bizarre glimpse into such a proto-diploid world thanks to some spoilsport genes which manage, even today, to subvert Mendel's rules. In principle, each chromosome should have the same chance as its counterpart to board a gamete: such a *fair draw* at segregation ensures that each gene is tested against a maximal diversity of background.* But some chromosomes learn to cheat, and to always come out on top after the shuffle.

There seem to be several tricks to this trade. One of the first examples was uncovered by Yuichiro Hiraizumi in 1956. It is played by the so-called *sd*-gene (*sd* for *segregation distortion*), together with some accomplices. The action takes place on chromosome II of the male fruit fly *Drosophila*. The time: that critical moment just before the two chromosomes II inherited from the fly's parents part company, after a long and fruitflyful cooperation, and board distinct sperm cells for their journey towards the egg.

The biochemistry of what is about to happen is full of twists and not yet understood in full detail. Let us simplify it to a Bugs Bunny-like cartoon which hopefully will not stray too far from the truth. Imagine that the two chromosomes drift in a balloon over a mountain range. Once they have crossed it and landed safely, they can go their separate ways (but toward the same goal). For the moment, they still belong to the same diploid fly body. But soon, each will be in a haploid sperm of its own.

One of them might hit upon a clever trick. Instead of waiting for the balloon slowly to descend, it leaves by parachute. Jumping is riskier than landing with the balloon, of course. But it pays to parachute, as long as the odds of surviving are reduced by less than half. Indeed, after a safe landing the jumper is sure to reach the egg first, while if it had stayed with the rival in the balloon, its chances of winning the subsequent race would only be 50 per cent. So, jumping is better. Hence, the jumping gene spreads. Finally, almost everybody will jump. Any advantage of parachuting is then gone: chromosomes are equally armed, once again. The probability of breaking one's neck has increased, that's all.

This Jumper gene is already a segregation distorter; it pays little regard to increase in fitness and the Fundamental Theorem. But the segregation distortion in *Drosophila*'s chromosome II works in a much more roundabout way. The Jumper gene does something active; now comes another gene which *sabotages* this activity.

It starts with a terrible error in one of the letters of instruction. Instead of *hold tight to parachutes*, the distorted message reads *hold light to parachutes*. So the parachutes are burnt. Since the next command says *now jump!* this mutation eliminates itself, together with the other passenger. Still, the mistake is likely to be repeated from time to time, until it occurs, by chance, in one of those few remaining chromosomes which do *not* jump. This chromosome survives: moreover, having burned not only its own parachute, but also that of its fellow passenger, who belongs to the jumping majority, the mutant will, after safely landing with the balloon, have the egg for itself. This is an immense selective advantage. The new chromosomal type (which burns the parachutes and does not jump) spreads at the expense of the old type (which does not burn parachutes, but jumps).

By cross-over, two other types will also arise. One burns and jumps; it promptly commits suicide. The other type burns nothing and refrains from jumping; this one may safely reach ground, even if its fellow passenger burns parachutes. Not having played with fire, its chances of remaining unharmed are actually better than those of the pyromaniac. In a population consisting mostly of non-jumping incendiaries, the frequency of non-jumping non-incendiaries will therefore grow. The more common they are, the better for them, since their chances of avoiding accidents will increase. After some time, the incendiaries have almost entirely vanished. Parachutes now remain mostly intact. That's when it becomes advantageous to jump. More and more chromosomes will be of the original type again: they take care of the parachutes and use them, too. But this is a golden opportunity for the tiny minority of non-jumping incendiaries who have survived. They spread like wildfire. And so, the next round is under way.

This game has a similar structure to that of Stone–Scissors–Paper: the Non-incendiaries/Non-jumpers are beaten by the Non-incendiaries/Jumpers, which are beaten by the Incendiaries/Non-jumpers, which in turn are beaten by the Non-incendiaries/Non-jumpers. Depending on the details of the model, the game will either end up in a stable equilibrium of the three chromosome frequencies (this is the case with the fruitfly), or it will lead to more and more violent upheavals. But in any case, such outlaw genes will cause average fitness to drop, by increasing the frequency of broken necks. The population may reach a genetic *dead end*, and possibly head for extinction.

It is obvious that a genome stands much to gain if it can suppress this sort of cheating. But that's not easy. David Haig (now at Harvard) and Alan Grafen from Oxford University, have shown that it would suffice to unlink the Incendiary/Non-incendiary locus from the Jumper/Non-jumper locus; for then, the fatal order to burn the parachutes would exterminate itself. The command for such an uncoupling of two instruction sheets would have to come from a third locus, of course. However, if this *cross-over* command is itself closely linked with the instruction to burn the chutes, it cannot spread: for then, it is in its interest (metaphorically speaking) that its chromosome *knows* that it should not jump. The command to *cross-over* is most efficient if it comes from another chromosome, where it has no vested interest and no distorted view. Haig and Grafen suggest that the partitioning of genetic instructions into several chromosomes is a defence against outlaw genes.*

Keeping it in the family

All human societies appear to have outlawed incest; it is the most universal taboo. In a curious reversal, however, some cultures have imposed close forms of inbreeding upon their ruling dynasties. The fear surrounding archaic taboos can still be felt in many myths and tragedy plays whose plots revolve around incestuous relations.

From the genetic point of view, incest deserves its bad reputation.* The main reason is that each one of us carries deleterious and even lethal alleles on several loci, which do not get expressed because they are recessive and occur in one copy only. Each such allele is rare: the probability that any one of them ever meets its like in a fusion of two gamete cells is extremely small.

But inbreeding increases this risk by a considerable margin. Take, for example, some such gene *a* which I have inherited from my father (Fig. 5.4). The probability that a (hypothetical) sister of mine has also received it is one-half. The (hypothetical) offspring of our incestuous union would inherit my *a* gene with probability one-half, and with the same probability my sister's *a* gene, if she has one. So, all together, the child inherits a double dose of *a* with probability one-eighth—which is a considerable risk. Every mating between relatives has a similar, if mostly weaker, effect.

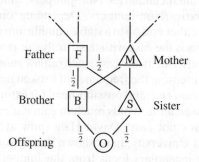

Fig. 5.4 The pedigree of the issue of a brother–sister mating. The numbers denote the probabilities that a rare allele present in one copy in the father (but not in the mother) of the incestuous pair will get transmitted.

This can be used for a definition of the *degree of relatedness* between two individuals: the probability that a very rare gene, if carried by one, is also carried by the other.* The degree of inbreeding of an individual is the degree of relatedness of its parents. The higher it is, the higher the overlap in its two sets of chromosomes. The recessive genes will then have to come to light.

Not all recessive genes are deleterious—far from it. But most deleterious genes are recessive. The reason is obvious. A gene which is dominant immediately exposes itself to the sharp wind of selection. If it is deleterious, it will be blown right off the map. But a recessive gene can thrive in the lee of its dominant partner.

Breeders must have noticed the detrimental effects of inbreeding at a very early stage. But the origin of the incest taboo is certainly much older. Is it innate? Behaviour which prevents matings between close relatives seems to have been observed in many species.* A genetic predisposition along these lines would be favourable. But of course, nobody has found a 'gene' against incest. It is conceivable that in humans, such a device has long since been made redundant by social constraints.

There is one form of incest which is out of our reach: it is self-fertilization. But plants do it all the time. If inbreeding is bad, self-fertilization must be worse, since father and mother are now not just related, but even identical. Hence we may expect that natural selection will do something about it. The safest recipe against self-fertilization is certainly to have separate sexes. But even hermaphrodites often avoid having offspring to whom they are father and mother in one. Some plants, like clover, have genes for self-sterility. They may use, for instance, a specific locus to probe the incoming pollen.* Like the daring suitor in the fairy tale, the pollen is asked a question. But unlike the fairy tale, the plant lets the pollen proceed only if it does *not* know, or more precisely if its answer sounds *foreign*; that is, if the pollen's allele is not of a type which has been inherited by the plant and transmitted to its eggs. It works like a negative password. The rarer a given answer in the population, the more frequently it finds open doors. No wonder that to some of these questions, many hundreds of admissible answers are known.

The selfing gene

There are, as we see, efficient ways of ensuring outcrossing; nevertheless, these devices are frequently given up. How can this be? A species yielding to the temptation of self-fertilization seems to commit a genetic folly. The offspring of a selfing plant will have only half as many loci bearing two different answers as its parent, since these answers will now stem with probability of one-half from the same parental questionnaire. After 10 generations of self-fertilization, the number of loci equipped with two distinct alleles is reduced by a factor of 1000. Soon every offspring will have a single answer to each question. Whatever the advantage of the genetic double game, here it is lost. In fact, the variety is reduced as fast as with blending inheritance. Nevertheless, many plants reproduce predominantly by self-fertilization (and not just weedy characters, either, but respectable soy beans, barley, or wheat).

The first to explain this was R. A. Fisher.* He performed his thought experiment on a vegetable hermaphrodite which does not fertilize itself. For simplicity, let us assume that its population is at a steady equilibrium: zero-growth. Since each plant has received two gametes, it will transmit only two gametes in its turn; on average, one egg and one pollen, since things have to balance out. Of course, each plant produces more than one egg and far more than one pollen, but so will all other plants. None can expect to pass on more than one egg and one pollen to the next generation.

But now, without any respect for decency, a freak turns up and pollinates itself: it provides each of its egg cells with a pollen of its own. These pollen are, of course, only a negligible fraction of its total pollen production. The remainder will be sown in the wind, like all other plants do. Again, one of those pollen will be successful (on average); and similarly, one of its eggs will make it too. But with this egg—and here is the difference—the freak plant smuggles an extra pollen of his into the next generation. It is as if, instead of drawing one card from the stack, one managed to take several. Altogether, with selfing, *three* gametes of the hermaphrodite make it to the next generation, rather than two. The policy of self-fertilization offers a huge advantage, therefore, and will sweep through the whole population.

At a closer look, the advantage shrinks down to size. The self-fertilized cell is not as viable as one produced by outcrossing. If its survival chance is reduced by half, for instance, then the trick of furnishing one's egg with a sperm of one's own offers no advantage any more. If the chance is three-quarters, however, then the bad habit will catch on. In this case, a trait will spread although it harms the population genetically, maybe to the point of extinction.

Even if extinction does not occur, the advantage of transmitting three gametes will not last long: for if in the end all plants fertilize themselves, then their free pollen will no longer find any vacant eggs, so that once again, only two gametes are passed on. In contrast to former times, however, the two gametes are now united in an organism of reduced viability, instead of trying their luck separately in two healthier offspring.

So the whole upheaval from outcrossing to self-fertilization has brought nothing but an increased risk of hereditary defects. A self-inflicted penalty, which just shows how silly selection can be. It rolls blindly on, inexorably like a Greek tragedy where fate is bent on steering towards incest.

It has to be said that it is not only the advantage of self-fertilization which decreases in time, but also its disadvantage. Since the deleterious genes cannot hide any more, they get bounced. The whole population is threatened with extinction; but if it manages to persist, then its hereditary material will have been greatly sifted. Once the population has gone through such an ordeal, further selfing becomes less dangerous. But of course, environmental conditions keep changing, and a population which has lost its genetic diversity has lost its faculty for adaptation.

In effect, to fertilize oneself is like giving up sex altogether: it seems difficult, for a higher organism, to keep this up for long. To quote Darwin, 'in no case, as I suspect, can self-fertilization go on for perpetuity'.* On the other hand, a short stretch of self-fertilization is easier to endure for those populations having only a few bad recessive genes to start with. This includes populations where inbreeding is frequent, for instance because their density is very low. With inbreeding, the bad recessive genes have been smoked out already.

Unmasking the rat queen

Plant and animal breeders can improve the genomes of their stock by submitting it to a few generations of inbreeding (to eliminate bad genes) and then crossing it with other inbred lines (to restore genetic diversity).

Interestingly, some natural populations prescribe themselves similar eugenic cures: termites, for instance. The founding pair of a termite colony inaugurates a dynasty even more devoted to incest than the Egyptian pharaohs of old. If the termite queen dies, for instance, then a daughter or a sister takes her place. The other offspring remain sterile members of the worker's caste. At some times, however, termites swarm. A strange rule of *Ladies first* ensures that females are allowed a few hours' head start. If male and female subsequently meet to launch a new termite stock, it will be almost certain that they are unrelated. After this outcrossing, a new round of intensive inbreeding begins (Fig. 5.5). As a result, deleterious recessive genes should be relatively rare among termites.

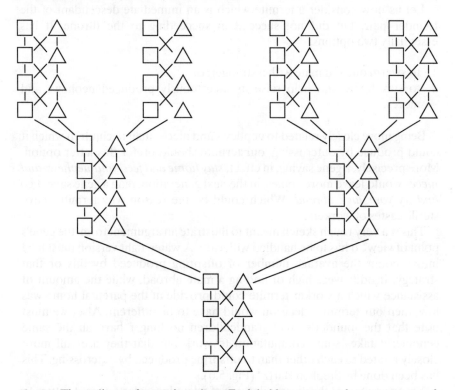

Fig. 5.5 The pedigree of a termite colony. Each ladder corresponds to a sequence of brother–sister matings within a colony. Their highly inbred descendants then swarm out and start new colonies by outcrossing with unrelated mates.

The alternation of outcrossing and incest leads to strange family ties. Each of the swarming termites is highly inbred, and therefore provided at most loci with a double dose of the same allele. Different termite stocks, however, will have ended up with different alleles; there is often no reason why one allele should be preferred to another, and its fixation in an inbred population is a random event.

It is not surprising that when the termite colony has reached a certain age, almost all of its members are genetically (almost) exact replicas of each other. The many rounds of inbreeding have seen to it. But what is remarkable is that the same holds for the *immediate offspring* of a colony's founder pair, which, as you will recall, is the issue of an outcrossing. This holds because both their father and their mother have only one answer to give to each question— although it will not necessarily be the same answer. It follows that the founder's grandchildren are issued from parents with identical genomes: they could just as well have been produced by self-fertilization—from one of their parents or, for that matter, one of their uncles or aunts.

Let us now consider a termite which is an immediate descendant of the founder pair, but did not succeed in succeeding to the throne. It has essentially two options:

1. to swarm out and mate with a stranger; or
2. to stay home and help raise its incestuously produced nephews and nieces.

Being more closely related to nephews and nieces than to children which it could produce by outcrossing, our termite should prefer this latter option. More precisely, a gene saying, in effect, *stay home and feed your nephews and nieces* would leave more copies, in the next generation, than a gene saying *go and try your luck abroad*. Which could be the reason why termites have sterile castes of workers.

This is a very rough sketch meant to illustrate an argument from 'the gene's point of view'. It has to be handled with care. A water-tight version must take into account the average number of offspring produced by this or that strategy. If odds were high of finding a mate abroad, while the amount of assistance which a worker termite could provide at the parental home was low, then our termite's decision would have to be different. Also, we must note that the founder's great-grandchildren no longer have all the same genome: it takes some computation to work out that they are still more closely related to each other than to offspring produced by outcrossing. This has been done by Stephen Bartz* (Fig. 5.6).

Yet an analysis of the diverse degrees of relatedness alone cannot suffice to explain sterile castes. Why shouldn't, for instance, a brother and a sister worker simply take off, hand in hand, and found their own colony?

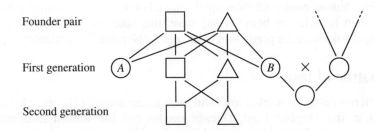

Founder pair

First generation

Second generation

Fig. 5.6 To stay or not to stay in one's termite colony? *A* and *B* are offspring of the founding pair, but did not succeed to the throne. *A* remains in the colony and helps raise the offspring of its luckier siblings. *B* leaves and finds an unrelated mate. Because their parents are highly inbred, *A* is more closely related to its nephews and nieces than *B* is to its own children.

Obviously because this cannot be as easy as it sounds. *Ecological opportunities* must play a decisive role. Apparently they are rare.

The British biologist W. D. Hamilton suggests that termites originated as species living under bark, digesting rotten wood.* Exploiting a dead tree trunk would be work for many generations, and offer a safe haven. But once the rotten wood is all consumed, individuals would have to swarm out and look for a new supply. The families would, by turns, be confined to cramped quarters and obliged to disperse. This could bring about a pattern of inbreeding and outcrossing which facilitates the evolution of sterile helpers.

This ecological aspect led to a most remarkable discovery* when, in 1976, the American biologist R. D. Alexander lectured on sterile castes. It was well known that these existed for ants, bees, and termites, but not for any kind of vertebrate. Alexander, in a kind of thought experiment, toyed with the notion of a mammal able to evolve a sterile caste. It would, like the termites, need an expandable nest allowing for an ample food supply and providing shelter from predators. For reasons of size, an underbark location was no good. But underground *burrows* replete with large tubers would fit the bill perfectly. The climate should be tropical; the soil (more than a hint of Sherlock Holmes here!) heavy clay. An ingenious exercise in armchair ecology, altogether. But after his lecture, Alexander was told that his hypothetical beast did indeed live in Africa; it was the naked mole rat, a small rodent studied by Jennifer Jarvis.

'Let theory guide your observation': the hypothesis that here, indeed, existed a mammal with sterile workers was quickly confirmed, after that. The naked mole rat fulfilled all expectations. Its colonies consist of one reproductive 'queen', a few male mates, and some hundred sterile workers of both sexes. The one snag is that the underground diggers, while highly inbred, as they should be, appear to show no stage of obligate outcrossing. But this missing element could still show up: the naked mole rats live at a much slower pace than termites or bees. In any case, this splendid feat of theoretical biology ranks with the prediction of the planet Neptune by astronomers.

A matter of taste

The sterile castes of worker ants and bees led to some of the most brilliant pages in the *Origin*.* First, Darwin singles out the 'almost insuperable difficulty' of explaining 'neuters [which] often differ widely in instinct from both males and fertile females, and yet, from being sterile, ... cannot propagate their kind ... so that [they] could never have transmitted successively acquired modifications of structure or instinct to [their] progeny.'

Next, Darwin offers what, as Alexander emphasizes,* is essentially the full solution. It is based on two simple ideas, both of which were well known to breeders at the time. One is that a trait can be transmitted without being expressed. This is shown by characters which reappear after having been lost for generations. It is shown by roosters being able to transmit factors for a higher egg production. It is also shown by 'oxen of certain breeds having longer horns than in other breeds, in comparison with the horns of the bulls or cows of these same breeds'. The oxen have no direct descendants, of course; and yet, they may be subjected to a selection specifically directed at them rather than the fertile members of their breed. The theme of sterility has been discreetly introduced.

The second of Darwin's remarks is 'that selection may be applied to the family, as well as to the individual... Thus, a well-flavoured vegetable is cooked, and the individual is destroyed; but the horticulturist sows seeds of the same stock, and confidently expects to get nearly the same variety.'

Darwin goes on: 'Thus I believe it has been with social insects: a slight modification in structure, or instinct, correlated with the sterile condition ... has been advantageous ... consequently the [related] fertile males and females of the same community flourished, and transmitted to their fertile offspring a tendency to produce sterile members having the same modification.' And then Darwin turns the table 'against the well-known doctrine of Lamarck ... For no amount of exercise, or habit, or volition, in the utterly sterile members of a community could possibly have affected the structure or instincts of the fertile members, which alone leave descendants.'

So sterile castes are no embarrassment at all. On the contrary: '... I should never have anticipated that natural selection could have been efficient in so high a degree, had not the case of these neuter insects convinced me of the fact.'

For Darwin's intellectual descendants, all that remained to be done was to quantify his argument. This started when Fisher explained the spread of bright warning colours among distasteful caterpillars, which might at first glance appear as a suicidal advertisement.* But a bird who swallows such a caterpillar will never do it again. If this saves the life of *more than two* siblings of the defunct—as will most probably be the case, since caterpillars travel in family groups—then the victim's death has not been in vain: more precisely, its genes for bright colours will be routed through the surviving siblings, and spread.

In the 1960s, W. D. Hamilton greatly expanded this reasoning and applied it to social insects. His 'three-quarters' rule* has become famous. The rule works because in bees and ants, females are diploid while males are haploid. (So drones have no father, but a grandfather.) A rare gene carried by a male will be passed on to all its daughters (instead of only half, if the male had been diploid). Hence sisters are more closely related than usual: their degree of relatedness is three-quarters instead of one-half (Fig. 5.7). Brothers, on the other hand, are not more closely related to their siblings than in 'normal' species (which include termites and naked mole rats). A female bee has more 'interest' in her siblings than a male, and hence a stronger incentive to help her mother. This strange asymmetry may explain why the sterile workers are invariably females in ants and in bees, while they are of both sexes anywhere else.

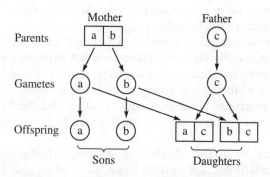

Fig. 5.7 A bee family: females are diploid and males haploid. A rare allele present in one copy in the queen will be transmitted with a probability of one-half to her daughters and sons. A rare allele in her consort, however, will never reach the son but certainly will reach the daughter. Daughters are therefore more closely related to each other than to their parents or to their own offspring.

The parliament of despots

The gene's point of view has been championed with enormous success in Richard Dawkins' book *The selfish gene*, which is an elaboration of R. A. Fisher's grand vision: the drama of natural selection acted out at the level of the genes.

It is no longer easy, nowadays, to appreciate the boldness of Fisher's step, but it crossed an abyss: on one side, well-specified yes-or-no characters like polydactyly (six-fingeredness) in humans and the properties of the peas studied by Mendel; on the other side, continuously varying traits like wing spans or hormonal levels. Artificial and natural selection usually act on traits of the latter type. Their slight variations provide the basis for evolution by continuous adaptation. On the level of the genes, however, variation is in steps, since one either has an allele (in one or two copies) or one has not.

If a character is specified by many genetic loci—where many means dozens, maybe even hundreds—then each gene's contribution is likely to be slight and will account for only a small step. But a transition in many small steps looks like a smooth transition. We know this from compact discs, for instance, where digital information (a sequence of integer numbers) can yield the smooth sounds of a string quartet.

What Fisher undertook as a young man was to prove that family resemblances, or more precisely the *correlations* between smoothly distributed traits in related individuals, agree with the Mendelian rules for the transmission of discrete factors.* Depending on the degree of relatedness, individuals share genes to a greater or lesser degree and should be more or less similar: this can help in estimating the relative roles of heritability and environmental effects—nature versus nurture—and to predict how a population responds to selection, for instance in a breeding program.*

No one can reasonably doubt nowadays that particulate entities govern the transmission of continuous traits. Genes are not hypothetical factors any longer. They can be seen, handled, spliced, and mutilated. And yet, in spite of brave efforts in mathematical genetics, the bridges betwen Mendel's rules and statistical assertions about continuous traits are fragile, to say the least, and hang on many simplifications—sometimes, in fact, they almost seem to float.

Let us assume, for instance, that body size is subject to stabilizing selection, which means that fitness is highest for some intermediate size. Let us assume furthermore that this size is influenced to an equal degree by each of some 20 genetic loci, which seems a conservative estimate. Each locus can carry two alleles that code for a fraction of a centimetre more or less. It is best, then, to have equally many + and − alleles. Which alleles will be favoured now?

It depends on the other genes, of course. This background is different for each individual. The value of a gene can be compared with that of a chess figure: it depends on the particular game just considered. In an end game, for instance, a well-placed pawn can be more important than a far-off queen. Nevertheless, we may safely assert that a queen has a higher value than a pawn—we just have to average over many end games. Similarly, we can define the value of an allele, or its contribution to fitness, by averaging over all possible gene combinations, each counted with its proper frequency. What is troubling, however, is that in our example this contribution is not likely to be significant at all. From the point of view of the gene, it does not really seem to matter which allele is present.* In such a situation, the gene can no longer be seen as the unit of selection.

Which leads to the question: what then—if anything—is selected? Is it the so-called *integrated gene complex*? Is it the *genetic cohesion*? Or is it *genetic homeostasis*, otherwise known as *coadapted harmony of the gene pool*? The artfulness of these expressions should make one's mental alarm bells ring. Does the coadapted gene complex sit on a single chromosome? If yes, what brought it there? Verbal reasoning is utterly unable to provide an answer. But so, alas, is mathematical analysis.

'The Mendelian was apt to compare the genetic contents of a population to a bag full of colored beans', to quote Ernst Mayr.* 'To consider genes as independent units is meaningless from the physiological as well as the evolutionary viewpoints'. J. B. S. Haldane was quick to take up the cause. His article 'In defense of beanbag genetics' turned the beanbag into a rallying point for population geneticists.* But this does not alter the fact that genes interact in horrendously complex ways. Mendel was incredibly lucky in choosing his traits, less because they were assorting independently, than because each depended on a single locus only. It is difficult to find similarly well-behaved traits, for instance in humans. An obvious candidate is eye-colour, which usually is inherited as if *blue* were recessive, but this, alas!, is not invariably so. Most human characters depending on one locus are either very rare (like albinism), or hard to detect (like blood group), or downright bizarre (like the ability to curl one's tongue or to taste the chemical PCT). There is no one-to-one correspondence, in general, between traits and genes. And further complications abound: some characters are only expressed in females, or only dominant in males (for instance baldness), or transmitted by X chromosomes which can never be carried by both father and son; some traits are highly dependent on their environment, or on maternal effects, or on variable expressivity or incomplete 'penetrance' of genes. For example, the son of Jacob Ruhe had that extra finger only on his left hand; on his right hand, he had a wart in place of the sixth finger—a watered-down manifestation of his gene which might misleadingly have served as an argument for blending inheritance. And the recent technique of *transgenics*—the transplantation of genes—has shown that in many cases it does matter, after all, whether an allele is inherited from the father or from the mother. When sex cells are formed, some genes can be altered by parental imprinting.*

Genes can have multiple effects, or interfere with each other. They modify, suppress, or regulate. Their chatter exceeds that of a poultry-yard; the *parliament of genes* is a favoured expression in this context. But parliament, and poultry-yard, are fairly transparent institutions. The genome seems rather like a nightmarishly huge administration, with lots of magistrates who do not know each other and do not wish to know, with countless departments, chambers, committees, and forgotten archives, each office filled with pig-headed clerks and secret rebels, idlers, and intriguers, a *bouillon* of red tape seething with nepotism, deceit, and obstinacy. Amazingly, however, this Kafkaesque organization works tolerably well in many cases.

The theoretical physicist Freeman Dyson has described the fuzzy regiment of genes by words which used to characterize the dual monarchy of the Habsburgs: *despotism, tempered by sloppiness.** One cannot express it any better. Perhaps it is not sheer coincidence that Mendel spent his whole life in this empire.

6

Evolution and sex

Sexual selection and the peacock's tail

The *good genes* theory

Bright plumage and parasites

Fisher's *runaway* process

Female choice and male competition

The origin of the male–female dimorphism

The prevalence of the 1 : 1 sex ratio and Fisher's explanation

Extraordinary sex ratios and local mate competition

Why two sexes instead of three or one

What use is sex?

Parthenogenesis and the cost of outcrossing

The Fisher–Muller theory and the Kirkpatrick–Jenkins theory on the incorporation of good genes

Muller's ratchet and the elimination of bad genes

The plausibility of group selection

The role of sex in speeding up evolution

The Tangled Bank, the Red Queen, and sib competition

The contrariness of the environment

Sex as a means for escaping parasites

6

Evolution and sex
The Only Game in Town

And why not? We all are gambling nowadays.
Fritz Lang, *Dr. Mabuse, the gambler*

The tell-tale tail

The peacock's tail outshines all its other endowments. It almost looks as if the bird wanted to cause a sensation at any cost. And the cost of the tail is high indeed: its aerodynamics are awkward and its brightness conspicuous.

The peacock has only itself to blame for this extravagance. If a lap-dog were reduced to living on its own, it would also have to pay dearly for its cute appearance. But the lap-dog is the outcome of artificial selection and was meant to spend a sheltered indoor life with its owner. Most peacocks, on the other hand, do still roam freely in the jungle. Their tails, which have earned them a distinguished place in parks and gardens as well as in numerous books on evolutionary biology, were not due to the fancy of a breeder but to a so-called whim of nature. Doubtless, many biologists and castle owners feel that if there were no peacock, it should be invented; but they have had no part in bringing it about. It was done by natural selection.

All attempts to ascribe to the king-size tail an advantage for survival seem rather artificial, however. It is unlikely, for instance, that the eye-pattern frightens predators away. Darwin was aiming for a better explanation. Although he used, after some hesitation, the expression *struggle for survival* (which was originally due to Huxley), he knew that survival is itself but a means to an end, namely propagation. A reduced probability of survival can be more than made up for by an increased fecundity. If the long tail is irresistibly attractive to peahen, it can provide its owner with a large number of mates and a heightened number of offspring, even while reducing the male's life expectancy.

Darwin denoted this as *sexual selection*.* It is not opposed to natural selection, but rather one of its aspects. But there are two opposite forces acting upon the evolution of the tail. *Lengthening it* means more success at mating. *Shortening it* means a better chance of reaching the mating season. The actual length is a compromise. Sexual selection can therefore reduce the probability of survival, and resembles artificial selection in this respect, with the peahen in the role of the breeder providing, for generation after generation, males with longer tails with a higher than average share of offspring.

Darwin's explanation has much in its favour: foremost, of course, the fact that only pea*cocks* have an oversized tail. If it offered an advantage for survival, it would be hard to see why females should not have it. The fact that successful peacocks have several mates also fits in rather well. It means some other peacocks have no mates at all. Whatever qualities they can boast of, they will not be able to pass them on. It pays to be sexy. While peahen can be fairly sure to end up in matrimony (even if it were only as a fifth concubine), peacocks have to display some special appeal to be a success with the opposite sex.

So it was female choice which caused the males' long tails. But what caused the female preference? Darwin simply took it for granted. R. A. Fisher offered an explanation in a few terse lines and a catchword: *runaway process*. Fisher was notorious for his laconic style. When his editor dared complain about it, he received for reply that 'fairly large print is a real antidote to stiff reading'.* (Fisher was, in fact, extremely myopic. His shortsightedness prevented him from taking up a career in astronomy; it also forced him to work things out in his head, and may thereby have caused his telegraphic style.)

It took theoretical biologists decades to understand Fisher's argument and check it by calculations; today, it is still subject to debate. This is not surprising, for according to the runaway theory the female preference was caused, essentially, by *itself*. This sounds somewhat like the tale of Baron Münchhausen, who got himself out of a bog by pulling on his own pigtail. Fisher's solution appears paradoxical at first, and requires some careful handling.

To start with, female preference has to be transmitted genetically. In the case of the peahen, we have to accept this on faith—it has not yet been proved by experiment. But a similar tendency for female ladybirds to prefer mates with a certain colouring has apparently been shown to be a heritable trait.* There are genes for female taste, then, just as there are genes for male tail length. It is probable that these genes are present both in males and females. They come into play only if the sex is right—otherwise, they are just carried along. (There is nothing strange in having peahen carry genes for a male peacock's tail: prize bulls, for instance, can transmit genes for a high milk yield.)

Next, the genes for the female taste must be promoted by natural selection. At first glance, the opposite seems to hold. A female accepting a suitor with a long tail will be likely to have sons with long tails and short life expectancy. There must be some advantage to compensate for this.

There is no lack of candidates, in fact. For instance, the peahen makes sure, by her choice, to mate with a true peacock, rather than with a similar bird.

Matings between distinct species can happen. If a female donkey, for instance, is mounted by a horse and gives birth to a mule, she expends a considerable amount of time and energy in a sterile *cul-de-sac* for her genes. It is obviously a sound policy to aim at avoiding such mistakes, and hence to develop clear signals and sensitive antennae for the recognition of suitable mates. In fact, species which look similar tend to differ most where their ranges overlap. As a clear signal, the spread tail of a peacock leaves nothing to be desired. But are peahens really so dumb that they cannot understand a less conspicuous signal? Surely a shorter tail should suffice as a membership badge to the peacock's club. The excessive nature of the trait remains unexplained, so far.

The best and brightest

The *good genes* theory approaches the problem from another angle. A female must have an interest in finding partners with a good genetic endowment, since her offspring will have good chances of inheriting it. But how can she gain information on the qualities of her suitors? The female fruitfly *Drosophila* tests the males by a swift sequence of dancing steps. If the male can follow suit, he is permitted to copulate; if not, then the female rejects the suitor. This trial by dancing makes sense: the clumsiness or exhaustion of the male can hint at a genetic defect. By rejecting his attempts at mating, the female avoids the risk of burdening her offspring with it. And the test works, as the British geneticist Linda Partridge has shown in a neat experiment.* *Drosophila* females permitted to choose had fitter offspring than females prevented from choosing, and could therefore 'expect' more grandchildren.

It seems likely that when a peacock raises its tail, it similarly displays vigour and health. This explanation has been expanded by William D. Hamilton and Marlene Zuk. Most species are permanently threatened with illness or debility caused by parasites, and must engage in a continuous arms race, their adaptive moves always being countered by moves from the parasite.* The female will obviously have an interest in finding for her offspring a father with the most efficient immune system currently on the market. But how can she check whether the suitor is plagued by parasites or not? Much like a poultry breeder would: by judging, for instance, the brightness of the plumage and the lustre of the eyes. Natural selection has no time to develop a specific sensor for this or that type of parasite, since the danger will come from another type in a few generations from now. The female is therefore well advised to look for *general* symptoms, like the state of the plumes, which can, of course, be judged best from a long tail.

Whoever buys a used car will usually take account of the brightness of its varnish. A prospective seller who does not even take the trouble to wash the car is not likely to have overhauled the motor or changed the brakes. To be sure, such a superficial judgement is less reliable than a careful examination by a competent mechanic. But the peahen has no diagnostic centre at hand, and a superficial check-up is better than none at all. After all, we also like to be told by acquaintances that we are 'in the pink'.

So if bright plumage promises a splendid state of health, this compensates (to some degree) for the increased danger from predation caused by the conspicuous tail. This danger falls under the heading of *advertisement costs*. The long feathers are a health certificate, according to this view. It is more than a clever speculation. Several investigations have confirmed that those bird species that suffer particularly from blood parasites tend to have an elaborate plumage.* Since it is not likely that parasites develop a fondness for brightly ornamented hosts, it must be the other way round: bird species which are particularly beset by parasites develop bright plumes. Female choice for healthy partners could provide the selective force.

Recently, Manfred Milinski and Theo Bakker have verified this in an elegant experiment, not on birds but on fish; more precisely, on stickleback, which are finger-long silvery freshwater fish whose lively disposition makes them a pleasure to watch. Male stickleback have a bright red belly;* but if they get infected by a ciliate parasite, they develop white spots which make their belly appear less red. Female stickleback prefer the redder males, and thus select for less infected mates. If their aquarium is illuminated by a green light, however, they stop caring one way or another, because they can no longer see the red. It must be the colour, therefore, which guides their choice, rather than the presence or absence of parasites as such.

The evidence is striking. *Good genes* makes good sense. And yet, there remains a lingering doubt whether this explanation can cover, so to speak, the peacock's tail in its entire length; it is perhaps too sensible to account for the glorious flamboyance of some ornaments. So let us turn to Fisher's runaway theory.

Runaway

A female's advantage in choosing males with long tails consists in having her sons inherit the sex appeal of their father. If the female predilection is widespread, it would be all wrong to swim against the tide of this fashion. But how do fashions spread? They are caused by reinforcement. Nothing succeeds (at least for a while) like success.

There are two sorts of feedback, positive and negative.* The negative feedback counteracts. It plays down: whatever is too large gets reduced, whatever is too small gets increased. This is used in countless steering devices and regulatory controls, from the very first steam engines onward. Positive feedback, on the other hand, exaggerates: whatever is large gets a further increase, and whatever is small gets still smaller. This can lead to a standstill or to an explosion, depending on which side of the threshold one starts from. A forest fire creating a storm, a crash at the stock-market or a melt-down in an atomic reactor are examples of such runaway processes. In technology and trade, positive feedback is often undesirable. In *advertisement*, however, it is aimed for; and wooing is, of course, a form of advertisement. (In German, the same word *Werbung* is used for courtship and publicity.) The exaggerated length of the peacock's tail indicates the work of positive feedback. But what sets it in motion?

A simple scenario runs as follows (Fig. 6.1). Let us begin with a population of birds where the females are supremely indifferent to the size of the male's

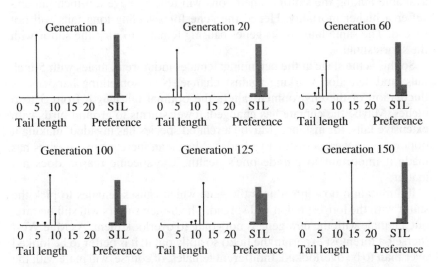

Fig. 6.1 A model for the runaway process. The population consists of 100 males and 100 females. Each individual inherits a gene for tail length (which can take the values 1, 2, 3, ...) and a gene for selecting male partners (which can be L, a preference for the longest tail; S, a preference for the shortest, or I, indifference). Depending on the individual's sex, one or the other gene operates. The probability of a male reaching adulthood is given by its tail length. For the currently optimal length, this probability equals one: for each unit more or less, it is reduced by 5 per cent. All females reach adulthood. Each female then surveys a randomly chosen sample of 10 adult males (its suitors); if the female carries gene I, it selects its mate at random, but if it carries gene L (or S), it chooses only among the males with the longest (or shortest) tail within the group of suitors. Each female produces one son and one daughter. Their genes are picked at random from the parent's genes. Once in a thousand births, a mutation can either change the length of the tail gene (by one unit), or alter the preference gene. The diagrams show the frequencies of the genes in the population. After 150 generations, the odds that a male survives to maturity are only 55 per cent. This run started off with a random fluctuation (it was not necessary to assume a shift in the optimal tail length due to an environmental change).

tail. This tail is subject to natural selection: the optimal length is, say, 5 inches. From time to time, genetic variation produces some birds with 4- or 6-inch tails, but these birds suffer a higher mortality rate.

If a mutation introduces females with a special taste for long tails, selection will tend to oust them from the population. Assume, for instance, that the mutant gene programs them to choose, among their suitors, the male with the largest tail. In a computer simulation, we could specify that each female enjoys the attentions of, say, 10 randomly chosen individuals from the adult male pool; she should look them over and elect a mate from among the suitors with the longest tails. The mutant gene for preferring long tails will cause its female bearers to select a 6-incher as mate, whenever one is available among the suitors. Their sons will tend to have a 6-inch tail and suffer a higher mortality. Hence the gene for selecting long tails will not spread. The same fate awaits genes causing females to pick the suitor with the shortest tail.

So this is the stage at the beginning: females indifferent, males with 5-inch tails, and selection working against change. Now something happens. A fluctuation of the environment alters the optimal tail length from 5 to 6 inches. Maybe the climate has changed: fewer storms to ground birds with extensive tails, for instance. Maybe a related species has invaded, making it opportune to wear a distinctive badge. Maybe an increase in parasites has made it important to parade one's health. The specific reason does not matter.

If a mutation now introduces the gene which causes females to pick the suitor with the longest tail, it will spread. The 6-inch suitors will still be rare, but a female with the new gene is not going to overlook them. As Mae West (herself a foremost expert in the field) so aptly put it, it is better to be looked over than to be overlooked. Indifferent females, of course, will mate with the optimal type only by chance: and this chance is initially very small, since the optimal type is so rare. Females with the preference gene will, in contrast, jump at every occasion available. And this time, they are betting on the right horse. They have more offspring in the next adult pool; these offspring will tend to carry the gene for the 6-inch tail, and—this is the essential point—the preference gene too, in spite of the fact that it only 'works' in females.

In the next generations, the frequency of 6-inchers will increase, and carry the genes for the female predilection along. Very soon, the frequency of these genes is a selective factor on its own: 6-inch males are favoured, not only by their lower mortality due to the change in environment, but by the action of that minority going for the longest tail. And this minority steadily grows.

Soon comes the stage where the tail-conscious females exert a stronger selective influence than the lower mortality does. In fact, the change in the optimal tail length, which triggered it all, has done its work by now and can depart: even if the environment changed back to the former conditions, and a

5-inch tail gave the better chances of surviving until maturity, it would no longer matter. Fashion dictates by now. So the 6-inch genes still spread, and with them, of course, the genes for preference.

Once there are a substantial number of 6-inch genes around, it becomes likely that one of them will mutate to 7 inches. Its chances of reaching adulthood are a little below average, but whenever its bearer competes for a female, it has a good chance of striking a responsive chord: which means more 7-inchers and yet more preference genes. And so on.

The genes for the female predilection form an alliance with the genes for the 7-inch tail in their long journey through the generations: it is fairly probable, indeed, that a male with a king-size tail has not only a father similarly endowed, but also a mother sensitive to its charm. When an 8-inch mutation enters the market, it will be eagerly welcomed by a group of females which is quite considerable by now, and the 7-inch model will find out that its alliance with female taste was only temporary. It will be given the cold shoulder whenever an 8-incher joins the party.

Shifting alliances all this time, the gene for the female preference spreads. It rides, like a surfer, on a wave of ever-increasing tail lengths sweeping through the population. In a way, the preference gene is allied to itself. To use an expression by Dawkins which has a biblical undertone, it 'recognizes' itself:* for it allows its female bearers to recognize males with long tails, and hence to find mates highly likely to bear a hidden gene which is a copy of itself. The genes for female preference form a *clique*, in this sense.

In this simple scenario, the gene for the female predilection may reach fixation. The males will respond by going to the very limit allowed for by mortality. It could also be that this limit is reached before *all* females bear the preference gene. In this case, there will be an uneasy equilibrium between preference and indifference in females, matched with an equally uneasy equilibrium between male tail lengths. Some simulation runs show, incidentally, that an initial advantage for the male trait is not necessary: a random fluctuation can set the process in motion.

If all males ended up with the same exaggerated tail length, the females could stop being choosy. But very likely, tail length is affected, not by one, but by many genetic loci. Thus there are many different ways of achieving the compromise length, and offspring will vary in tail size. This keeps selection on its toes, so to speak: the preference gene will always have something to do.

There are other, considerably more sophisticated models of sexual selection.* Our hypothetical gene for preferring the male with the longest tail can be replaced by a genetically programmed 'search image' for an ideal length, for instance. The sexually selected trait need not be confined to males, by the way. We notice traits more easily if they differ between the sexes, that's all.

Gift wrappings

But why are the females the choosers and the males the suitors? Actually, we know of cases where the roles are reversed: in quite a few species of birds, for instance, the females wear the brighter colours.* The selective breeding has in such cases been done by the males, who accordingly bear the incumbent costs (like constructing the nest, guarding the eggs, and raising the young). It is the females, here, who can afford territories with several nests. They may increase their reproductive success by practising polyandry—and for that, they have to be sexy.

But these are exceptions proving the rule. In the overwhelming majority of cases, be it birds, fish, mammals, insects, or spiders, the competition is between males. This competition can take the form of fights with rivals, or of display and courtship for the favour of the females. Both types of competition lead frequently to the development of special male traits which serve as ornaments or weapons (or both), like the antlers of a deer or the tusks of a walrus.

Male competition is a prime mover of evolution; not only of biological evolution, but of cultural evolution too. According to the Dutch historian Huizinga, play was a decisive factor in early civilization; and the two main factors of social play in archaic societies were 'glorious exhibitionism and agonistic aspiration'.* These are exactly the two faces of male competition.

But let us stick to biology and ask how it happens that in most cases, it is the females who are wooed and fought for? What makes them so valuable? They are, after all, not rarer than males. The sex ratio in most populations' is approximately 1:1 (which also begs for an explanation, by the way).

How do females differ from males, in the first place? The answer is easy: they have the larger gametes. This is, in fact, the definition of *female*. The gametes of males (the sperm) are tiny cells consisting of not much more than the genetic material and, often, a tail for propulsion. The gametes of the females (the eggs) are, on the other hand, well supplied with resources and therefore much larger. (The largest cells on Earth are eggs.) Females, accordingly, can produce only comparatively few gametes, and males enormously more.

It follows that males can afford to waste their sperm, while females have to *husband* their eggs. This holds even for insects. A female fruitfly produces only a few thousand eggs during her life-time, which can be fertilized by the sperm from one single mating. Small wonder that the female will be coy and try a few dance-steps to check the qualifications of her suitor: she is literally putting all her eggs into one basket. In contrast to this, male fruitflies produce many billions of sperm cells and have not developed any restraint in spreading these cheap mass products around. Maynard Smith describes how they will even attempt to copulate with blobs of wax of the right size.*

What caused the two sexes to package the genetic instructions for their offspring in such different ways? It seems to have been done by positive feed-back, again.* Let us suppose, indeed, that originally both sexes were producing gametes of the same number and size. Of course, they would not be *exactly* the same, but rather dispersed around a common mean. Whoever produced gametes which were too small was at risk: the corresponding zygotes (the first cells of the offspring, obtained by the fusion of a gamete with a gamete from the opposite sex) were liable to be under-equipped and hence unviable. Whoever produced larger gametes, on the other hand, produced fewer of them: this would also reduce one's offspring. Such deviations cannot spread. There is an optimal solution to the problem of parcelling a given total amount of resources into gametes.

But this solution will not be stable. At some time or other, a random fluctuation will cause one of the two sexes, say sex A, to have an average gamete size slightly above the optimal value. Whenever this happens, those of sex B producing gametes of less than average size are no longer penalized: the small amount of resources which they keep away from their offspring is provided for by their partner. In fact, far from being penalized, these pro-ducers of smaller-than-average parcels fare somewhat better than average: indeed, they produce *more* gametes, and these gametes have now good chances of surviving.

Will the disposition for producing smaller-than-average spread? Not necessarily. In the next generation, indeed, this disposition can find itself in a body of sex A. But suppose that the program for parcelling becomes conditional. It can spread if it says, in effect: *whenever you happen to be of sex B, make your bundles smaller and more numerous.* Such a conditional program is not implausible, as we shall see. The average size of gametes produced by sex B will then decrease. Within the other sex, the slight increase in gamete size originally due to chance will now be bolstered by natural selection; indeed, those who stick to the norm risk producing non-viable offspring, since their partner's gametes are likely to be undersized. There is no place to lodge a complaint. Better to adapt to the shifting conditions with a program saying *whenever you happen to be of sex A, make your bundles larger and less numerous.*

The random fluctuation has caused a trend. From now on, one sex has a larger gamete size than the other. There will be further random fluctuations, of course, but the symmetry between the two sexes is now broken. Under very general conditions, a deviation which reduces the difference between the sexes will be less favoured than a deviation which increases this dif-ference. This positive feedback is caused, not by a clique this time, but by a coalition of extremists. The females (we can now name the two sexes) produce larger and larger gametes, the males smaller and smaller, until a limit is reached. Since the females compete with each other, and likewise the males, the former cannot help being exploited, and the latter cannot afford not to exploit (Fig. 6.2). The split is irreversible, and deepens like a ravine cut by a torrent, until the size of the gametes offers no further scope for differentiation.

Other factors will step in and accentuate the divergence. The fact that it is invariably the female who bears the young in mammal species is just another

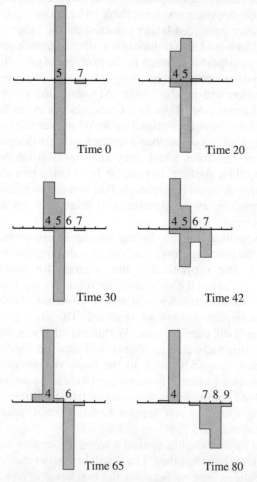

Fig. 6.2 Two mating types go their separate ways. The number of divisions of the total gametic mass is shown. Initially, each individual splits its gametic mass five times (that is, into $2^5 = 32$ gametes of equal size). A random fluctuation causes a few members of one mating type to divide 7 times. (In each generation, there are 100 individuals of each type. The total gametic mass is 500 for one type and 510 for the other. A zygote with mass less than 30 has no chance of surviving; with mass 34, the survival probability is 100 per cent; in between, the survival probability grows linearly. Mutation rate is 0.5 per cent.)

escalation of this fundamental inequality. Whichever sex has invested more than the other can ill afford to resist further exploitation. Its own loss in breaking up the joint venture is higher than the partner's. Life is not fair. The species where the female succeeds in getting the male to carry the main burdens of reproduction are a minority. In principle, the male pursues a parasitic strategy.

Y all those boys?

Females are more valuable, then. This makes them a bottleneck for population growth. Most males are superfluous. There are many species (for instance, humans) where the sperm of a single male would suffice for inseminating all eggs of all females. Why are there, then, so many males around, roughly as many as there are females, in most species? Every animal breeder knows that this ratio is not favourable, and makes sure of working with a considerable surplus of females. Why does natural selection not imitate, or rather anticipate, the breeder? It seems that the number of sons contesting for females is highly inflated, a waste of parental effort.

Again, it was R. A. Fisher who found an explanation—this time, a negative feedback.* If the proportion of females exceeds 50 per cent, it is advantageous to produce more sons; if the proportion of males is larger, on the other hand, then it pays more to have daughters. This regulates the sex ratio. It is like a party: if one sex is in excess, the other is in demand.

Before going into details, a clarification is called for. The familiar mechanism which determines the sex of our children cannot (yet) be affected by the parents' wishes. Females have two sex chromosomes of type X, males one X and one Y chromosome. Every egg contains one X, while sperm cells have with equal probability one X or one Y. The former lead—after fusion with an egg—to an XX offspring, and hence to a daughter; the latter, to an XY, and hence to a son. This mechanism holds for all mammals, and also for all birds (except that it is the *female* birds who are XY). At first glance, this seems to make a 1:1 sex ratio more or less inevitable. But this is actually not the case. The sex ratio *at conception* is usually male-biased, and only later balances out. For humans, at three months after conception, there are still 120 males to every 100 females.* Due to the higher mortality of male embryos in the uterus, this ratio falls off to something like 106:100 at birth (there is some puzzling geographical variation in the statistics). After birth, males keep being subjected to a higher mortality, so that at age 15–20, the ratio levels out: later in life, it is female-biased.

We shall return to this age-dependence of the sex-ratio in a moment. Right now, we use it only to conclude that the XY mechanism does not necessarily lead to 1:1. It seems plausible that if another sex ratio promised an appreciable selective advantage, nature would find a way of reaching it: for instance by different mortality rates of X and Y sperm. There exist, by the way, many other mechanisms for determining sex in the offspring: silverside, for instance (a small silvery fish very common in the Atlantic) has its sex partly determined by the temperature of the water at its birth. Low temperatures yield females and higher temperatures yield males. When averaged over the year, the sex ratio is 1:1.

Since there are so many different ways of influencing whether an individual is male or female, natural selection seems to have plenty of opportunities for expressing itself on the subject; if it did not favour a balanced sex ratio, it could probably tilt it.

Although individual parents have no choice on the matter, it will simplify Fisher's thought experiment to pretend that they do (Fig. 6.3). So let us suppose that the mother, say, which we are going to call Alpha, can determine the child's sex at will. The total number of Alpha's children is given, and does not depend on their sex. But the number of her *grandchildren* is affected by her choice, if the sex ratio within the population happens to deviate from 1:1. If it were 2:1, for instance—say 2000 males and 1000 females in the generation of Alpha's children—then Alpha would opt for a daughter. At least, she should do so if she could figure out the solution to the following exercise. If Gamma is a randomly chosen member of the third generation, what is the probability that it is a grandchild of Alpha (more precisely, an offspring of Alpha's child Beta whose sex is just being decided upon)?

Well, Gamma has one father and one mother. There are 2000 candidates for the role of the father. If Beta is going to be a son, its chance of becoming Gamma's father is one in 2000. If Beta is going to be a daughter, on the other hand, then its chance of becoming Gamma's mother is one in 1000. Hence Alpha doubles her odds of being Gamma's grandmother if her child Beta is a daughter. By choosing Beta's sex accordingly, Alpha can double, therefore, the number of her grandchildren. The same argument shows that conversely, if there is a surplus of females, it is better to produce a son. This is a situation where it is best to move *against* the crowd: hence, to invest in whatever sex happens to be in minority.

We started with the assumption that it is the mother who determines the sex of the offspring. It could just as well be the father, or the newborn itself. It doesn't matter, because this is a game where the interests of all three participants agree. Whatever the precise nature of the mechanism determining sex, natural selection will favour those factors which bring the average sex ratio in the population closer to 1:1.

Fish support Fisher

Fisher's sex ratio theory has recently been checked with silverside.* David O. Conover, a marine scientist at the State University of New York at Stony

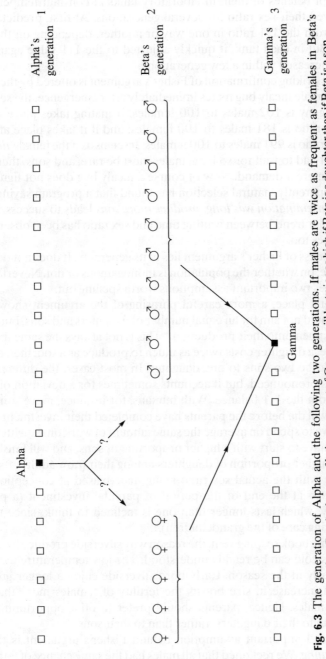

Fig. 6.3 The generation of Alpha and the following two generations. If males are twice as frequent as females in Beta's generation, then Beta's chance of being a parent of Gamma will be twice as high if Beta is a daughter than if Beta is a son.

Brook, kept batches of them in laboratory tanks at constant temperatures and followed their sex ratio for several generations. At first, predictably, it deviated from the 1:1 ratio in one way or another, depending on the temperature. But in each tank, it quickly adjusted to the 1:1 value again. The rarer sex increased within a few generations.

Another striking confirmation of Fisher's argument is offered by the mealy bug.* If a female mealy bug mates immediately after emergence, the sex ratio of her progeny is 102 males to 100 females. If mating takes place after 6 weeks, the ratio is 181 males to 100 females; and if it takes place after 10 weeks, the ratio is 991 males to 100 females. It seems as if the female *deduces* that since it had to wait for so long, males must be rare, and sons, therefore, would be in great demand. Now of course a mealy bug does not figure this out: but apparently, natural selection has found that a program saying *if the wait until insemination was long, produce more sons* leads to success. In any case, a similar trend between waiting time and sex ratio has been observed in other species too.

The neatness of Fisher's argument lies in its generality. It does not depend, for instance, on whether the population is monogamous or not. Nevertheless, it tacitly uses two important assumptions worth spelling out.

In the first place, a more careful phrasing of the argument shows that parents should not aim for an equal number of daughters and sons, but for an equal investment into their production.* This is not always the same thing. If, for instance, a daughter costs twice as much to produce as a son, then the sex ratio should be two sons to one daughter. In most cases, the difference in costs is less pronounced, but it accounts sometimes for a deviation of a few per cent from the 1:1 balance. With humans, for instance, more sons than daughters will die before the parents have completed their investment. Since parents have to spend on average the same amount of work on daughters and sons, they have to start with a higher proportion of sons, and will finish their job with a higher proportion of daughters among their (now adult) offspring. This fits in with the actual sex ratio being male-biased at conception and female-biased at the end of the period of parental investment (a period, incidentally, which lasts longer than one is inclined to think, since it may include taking care of the grandchildren).

In the same bookkeeping vein, the reason why silverside produce females if the water is cold can be readily understood. The low temperature indicates that it is early in the season. Early-born silverside enjoy a longer growing period. An increase in size boosts the fertility of females more than the fertility of males. Hence, parents should prefer to offer opportunities for extra growth to their daughters, rather than to their sons.

The second important assumption behind Fisher's argument is that of *random mating*. We reckoned that all males had the same chance of fathering Gamma. This need not be the case if pairings are assorted (as by an animal breeder, for instance), rather than random. Hamilton explained some extraordinary sex ratios in this way.* If, for instance, a parasite's offspring grow up in isolation—somewhere in the innards of a host, for example—then the young female parasites have no choice but to mate with their brothers. Their parents will, in this case, maximize the number of their grandchildren by producing only one son per batch.

Two is company

Behind the sex ratio, other questions are lurking. Why are there two sexes? Why not three, for example?

We can conceive a population with three sexes in several ways. There could be three different types of pairings—One with Two, Two with Three, and Three with One. But it is hardly likely that all three types of pairings had *exactly* the same success. Now if one type had an advantage, however slight, then it would tend to predominate. Whichever sex is left out in this pairing is doomed to vanish in due time.* (This explanation would seem more water-tight if there were not many plant species with several types of pairings, in fact. More precisely, such plants are divided into so-called *incompatibility types* which preclude certain pairings or render them infertile. This is basically a mechanism to prevent self-fertilization.* But these incompatibility types show that there could be some advantage for the establishment of a third sex. Whenever the third type becomes rare, it can mate with almost anybody else, which is a great bonus if time is limited.)

One could also think of a 'pairing' involving three sexes at once: One with Two *and* Three. In such a population, an offspring would have three parents. A mechanism for fusing three genomes would be rather com-plicated, to be sure, but certainly seems conceivable—and nature doesn't shrink from complications, especially not with sex-life. Nobody seems to have thought up an advantage for having three parents, but this does not prove that none exists. (And incidentally, one could argue that bees have three 'sexes'—queens, drones, and workers. The latter do not pass on their genes, at least not directly, but play an indispensable role in raising the offspring.)

The question why there are different sexes at all is still more difficult. Why does not one sex suffice? Why are gametes split into two classes, and prevented from fusing with members of their own class?

One theory assumes that—in contrast to the scenario described here—the differentiation into large and small gametes came first, and that subsequently, sperm cells were under desperate pressure to find a means for preventing their fusion with other sperm, since such a union must lead to under-equipped cells.* But this does not explain why two eggs do not fuse, either. Their union would be overweight, but so what? Surely it cannot do any harm to have plenty. Besides, there are species having two sexes both producing gametes of the same size.

It is possible that the separation into two sexes is meant to promote outcrossing. If organisms simply shed their gametes (into the surrounding water, say), self-fertilization would be likely to occur, with its attendant risk for the offspring's genome (see Chapter 5). Hence, it would be useful for the gametes to be provided with some device V preventing them from fusing with other gametes carrying V. Individuals with such gametes need no longer worry about self-fertilization. If obligate outcrossing is sufficiently advantageous, then the device V should spread, but of course not through the entire population. Some other gametes must remain. But these would still be plagued by the danger from self-fertilization. One of them could develop a similar device U (some chemical signal, perhaps). U-gametes which can fuse with V-gametes will be better off than U-gametes which cannot. The other types of gametes will be outcompeted, and what remains are the two sexes.

Oxford biologist Laurence Hurst has advanced a radically different explanation for having separate sexes.* As we have seen, genes need not work for the best of the genome, but may promote their own propagation at the expense of the collective's welfare, for instance by distorting the Mendelian 'fair draw' in their favour. Within the nucleus of a cell, this genetic tussle is somewhat muted by the fact that a genome needs a double set of chromosomes. But the cell resulting from the fusion of two gametes also inherits some genetic material which does not go into the nucleus, but is contained in the mitochondria (or other organelles). Remarkably, the cell does *not* need a double dose of these genes: one set of mitochondrial genes is enough. Which of the two gametes is going to provide for it? Genetic rivalry, here, is apt to become a fight to the death. It becomes essential to eliminate the opposite number.

In a gamete fusing with another, the genes in the mitochondria are all set for a desperate mêlée, therefore, while the genes in the nucleus are ready to engage in a domestic alliance which, while not free from contention, preserves a united front to the world. Hence nuclear and mitochondrial genes take different stands. Nuclear genes will tend to avoid getting embroiled in a fierce brawl. They have an interest in shutting down their warlike mitochondrial escorts. If two dog-owners wish to date each other, but their dogs are spoiling for a fight, then the owners will leave them at home. It is true that the nuclei from two gametes cannot *both* shed their mitochondrial genes, since they will need one set after their fusion. But one of the two gametes can relinquish its mitochondria in a kind of unilateral disarmament. Of course it then must make sure not to fuse with a gamete which has also surrendered its part. The other type of gamete will quickly catch on and fuse only with partners without mitochondria.

Being male, according to the new definition, means producing gametes containing only genes from the nucleus. This is a surprising about-face. According to the usual view, males are characterized as producers of small gametes: because the gametes are so small, they have no room for mitochondria. It could be the other way round. First, males omit to pack mitochondria into their gametes: once this asymmetry is established, it is bound to lead them to smaller and smaller sperm, thereby forcing the females to make up for it. It is an ironic tale. Males start out by courteously giving way on a small matter. They end up by exploiting females to the hilt.

According to Hurst and Hamilton,* if mating types manage to suppress warfare between mitochondrial genes, they constitute sexes; whereas if they only promote out-crossing, they are incompatibility types. Some organisms use both mechanisms. There are species of ciliates whose members can recombine their genes by either fusing or conjugating. If they conjugate, they only exchange genes from their nuclei: matings are arranged according to incompatibility type. But if they fuse, they also transmit non-nuclear genes: and then, gratifyingly, they have sexes.

The Dutch geneticist Ralph Hoekstra, one of the leading researchers on the evolution of sexes, points out that there has been, so far, a painful lack of experimental work on sexual differentiation.* This field has been neglected for too long. Hoekstra quotes R. A. Fisher: 'No practical biologist interested in sexual reproduction would be led to work out the detailed consequences experienced by organisms having three or more sexes; yet what else should he do if he wishes to understand why the sexes are, in fact, always two?* Some slime moulds actually have 13 sexes; more importantly, as Hoekstra points out, 'it seems remarkable that Fisher did not consider the possibility of analysing the logically simplest case of one sex'.

The luxury of keeping males

As many biologists have stressed, the very fact of sexual propagation is based on a paradox: propagation, indeed, means that one divides in two, while sex consists in fusing two cells to make one.* It appears abstruse to combine these two processes. More than that: it seems almost incomprehensible that anyone can afford it. If a female individual is able to produce offspring without being fertilized, then she enjoys a vast reproductive advantage over her female rivals. A gene allowing her to multiply this way would find itself in all her offspring. The offspring would consist of daughters inheriting the entire genome of their mother, rather than only half of it. Strictly from the genetic point of view, sex is horrendously costly.

This idea deserves to be run in slow motion. We are considering beings whose chromosomes occur in pairs, one from each parent. These pairs part company at *meiosis*, and each gamete contains only one chromosome of each pair. Offspring are produced by fusing two gametes, so that the chromosomes are paired again. But many beings can reproduce in another, *asexual* way, without meiosis and without fusion. Their gametes contain a full genome, which is—if no mutation has meddled with it—exactly the same as their parent's. Such a gamete has of course to be fully equipped with all necessary resources for survival, and hence must be an egg. The parent must be female, therefore. Her way of propagating 'like a virgin' is called *parthenogenetic*. It does not occur in mammals and freely living birds; but in fish, lizards, frogs, insects, spiders, worms, and plants, it is not uncommon.

Now let us think of a species where the male has no part in raising the young. In this case, a parthenogenetic female can rightly expect as many adult offspring as her sexually reproducing sisters. But the offspring of the parthenogenetic female consist of parthenogenetic daughters, with the same genes as their mother. By contrast, the offspring of a sexual female are with the same probability male or female and carry only half of their mother's genetic instructions.

The male, once again, is cast in the exploiter's role. Not only does he take no part in the production of valuable eggs; he uses them for passing his own genes along. They journey as stowaways to the next generation. Fertilized eggs contain only half of the genes of the sexually reproducing mother. How can she afford to dilute her genome to such a degree? Whenever a mutation produces, within a sexually reproducing population, a gene enabling the female to switch to parthenogenesis, this gene will double in frequency from one generation to the next (Fig. 6.4). From another point of view, one can say that the parthenogenetic mother invests exclusively in her like and skips the production of sons (Fig. 6.5).

This argument, which was first spelled out in 1971 by John Maynard

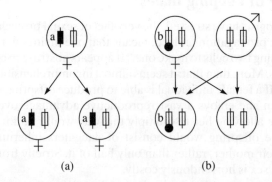

(a)   (b)

Fig. 6.4 The prospects of a new mutation (a) causing the female to reproduce parthenogenetically and (b) not altering her way of reproducing sexually. Mutation (a) finds itself in all offspring and hence spreads twice as fast as (b). (After Maynard Smith 1989.)

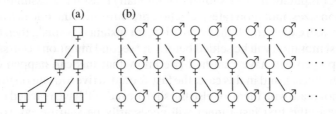

Fig. 6.5 Every mother provides two offspring for the next generation. (a) A parthenogenetic mother produces two parthenogenetic daughters; (b) a sexually reproducing mother produces one son and one daughter, on average.

Smith, is known as the 'twofold cost of sex',* or (more precisely) as the 'twofold cost of outcrossing'. The cost is essentially due to the needless production of male exploiters. If males, however, carry their full share of parental care, then both partners invest equally in their common offspring, and can be expected to raise twice as many as the single mother. This would offset the disadvantage, for a new gene, of finding itself with only 50 per cent probability in an offspring. It can try twice as often.

When sex first originated, there was probably no differentiation into males and females, and the cost of sex, therefore, did not have to be paid. This cost is a maintenance cost, not an admission cost. It is an age-old trick of exploiters to offer the first round for free.

Fast sex and clean sex

One of the first attempts to explain the evolutionary reason for sex is due to Fisher (of course). As usual with him, it more or less reduced to a laconic hint.* This was later worked out, in particular by Hermann J. Muller, an American geneticist from Thomas Hunt Morgan's *Drosophila* school who was awarded the Nobel prize for his work on mutations.*

According to Fisher and Muller, sexual reproduction enables a population to evolve faster, by rapidly bringing together two beneficial mutations. Such mutations occur only very rarely. Let us suppose that one improves eyesight and the other running speed. How can it come to individuals with sharper eyes *and* faster feet? In an asexual population, it can only happen if the second mutation occurs within the descendants of the first mutant. As long as its progeny is not widely spread, this is highly improbable. With sexual reproduction, on the other hand, such a lucky strike is not required. It is enough for descendants of the sharp-eyed to mate with descendants of the fleet-footed—an occurrence which is much more likely.

This argument, which held sway for several decades, has nowadays lost ground. Computations have shown that for many reasonable assumptions on population sizes, mutation rates, selection pressure, etc., the rate of evolution is slower than expected.* If the population, for instance, is small, then the fate of the first mutation will be settled before the second mutation occurs. Either the sharp-eyed will have vanished (we know that this can happen despite their high fitness), and in this case, the fast-footed arrive too late on the scene and cannot mate with them, or else the sharp-eyed will have reached fixation. In this case, the first fast runner will necessarily be sharp-eyed, so that a mating is no longer needed anyway. It is mostly in very large populations that sex speeds up evolution, especially if the mutations are of small advantage only and occur at many different spots in the genome.

In the 1950s, Muller pointed out that sexual reproduction acted not only to combine good genes, but also to eliminate bad genes.* In asexual populations, mutations pile up; it is highly unlikely that one mutation undoes a previous mutation. Now since most mutations are deleterious, genomes will carry an ever increasing load of harmful genes. To be sure, some genomes will be better off than others. The population will contain a privileged group of individuals carrying only the minimal load. This group should grow a bit faster than the others, since it is suffering less from adverse selection. But it is liable sometimes to be reduced, either because further mutations strike, or because, by a fluke of *random sampling*, less than its due share of offspring reaches the next generation. In fact, if the privileged class is small, this last effect is likely to wipe it out entirely. This is known as *Muller's ratchet.* Whenever the class with the currently minimal load is eliminated, the ratchet has clicked round one notch: the minimal load of bad genes has increased. This process can be repeated, but never reversed.

Such a gradual and seemingly inexorable accumulation of mutations can be offset by sexual reproduction. If an offspring is lucky, it can inherit mostly undamaged parts from its parents' genomes. Its load of bad genes is lowered in such a case. Sure enough, this works both ways. The offspring can also inherit an above-average share of bad genes, and have a load which is higher than its parents'. So it is hard to see what an individual can expect from the fact that its mother had sex. But for the population as a whole, the reshuffling of genes works like a health-care program: deleterious mutations are removed from some genomes, and accumulated in others which will be eliminated by selection. The American sociobiologist Robert Trivers compared sexual reproduction to a good broom: the dust is swept into one corner of the yard, ready to be shovelled away.*

This eliminating of bad mutations seems at first glance very different from combining good mutations, but Muller's ratchet is, in fact, closely related to the Fisher–Muller theory. In one case, the advantageous genes are supposed to be very rare, and in the other case, very frequent. In both cases, they are brought together by the sexual recombination of parental genomes. Most theoretical biologists agree that recombination is the most important aspect of sexual reproduction. As Muller wrote in 1932, *the essence of sex is Mendelian recombination.**

In principle, however, sex could work without recombination (if each gamete contained one chromosome only, and if this chromosome were proof against crossover). Recently, M. Kirkpatrick and C. Jenkins have proposed an argument in favour of sex which works without invoking recombination.* It describes the fate of a *single* type of advantageous mutation (causing, say, better eyesight). We shall assume that it becomes fully effective, however, only if carried by *both* chromosomes. Such dominance effects are extremely common. In an asexual population, descendants of a mutant having the new gene on one chromosome will have to mark time until a further mutation brings about another specimen of the new gene on the second chromosome. Only then can they have truly sharp eyes. Within a sexual population, on the other hand, there is no need for the second mutation. A small amount of inbreeding between the descendants of the original mutant can produce individuals inheriting the new gene both from their mother and from their father.

However, the idea of Kirkpatrick and Jenkins fails to explain the prevalence of outcrossing (since the effect works best with inbreeding), and it gives no clue as to why, in almost all higher organisms, the genome is split into several chromosomes freely recombining under sex.

The proper place for sex

It cannot be doubted that sex plays an important eugenic role in restructuring a species' gene pool. But can this be the reason it persists? Superficially, at least, this seems to ascribe a kind of foresight to natural selection. The good Darwinian frowns at that sort of thing. Should a mother spurn the immediate advantage offered by parthenogenesis just to provide her species with the means for reducing its load of mutations and for adapting at a faster rate? This looks like a clash of interest between the individual and the group. Such conflicts are frequent. They are usually settled in favour of the individual. It seems different here: group selection appears to be winning.

This outcome, to be sure, is not *unthinkable.** It could well be that a newly occurring hereditary program for parthenogenesis spreads at first like wildfire through the whole population (since it will be transmitted, at least initially, at a twofold rate), but that the population subsequently fades to extinction because it is unable to adapt to environmental change with the necessary dispatch, or is unable to brush off deleterious mutations.

Accordingly, only those populations would last which, by luck, are never tempted away from the path of sexual intercourse. Asexual species should tend to be short-lived: those which are present today should have shown up only recently. This appears to be the case, in fact. In the words of James Crow, asexual species form twigs of the phylogenetic descendency tree, rather than its main stems and branches.*

For most species, it seems difficult to acquire the knack of virgin propagation. Mammals, probably, have never been able to do it; and all the efforts of animal breeders have not yet succeeded in realizing the dazzling prospect of a poultry race consisting exclusively of egg-laying females. Furthermore, the parthenogenetic females of frog species still need male sperm for the activation of their eggs' development (but not for any supply of genetic material).* Within their reach, there must be closely related sexual species of frog whose males have not yet 'understood' that copulation with an asexual female is a case of, literally, love's labours lost: the offspring will not carry any genes from the misused male. This strongly suggests that the asexual frog line cannot be very old: the males, otherwise, would be more discriminating and the females more self-reliant.

Even such staunch critics of group selection as Fisher or Maynard Smith admit that it may have favoured the evolution of sex. But there must be more immediate advantages for sexual replication. After all, there are plenty of species of arthropods and higher plants that replicate exclusively by parthenogenesis. More remarkably still, some other species switch at the drop of a hat (or from one season to another, at least) from sexual to asexual multiplication and back again. These species, which ought certainly to be able, if pressed, to foresake sex forever, prefer to keep dallying with it. This is intriguing. The twofold advantage offered by giving up sex seems to be always at hand, and yet used only sparingly. Quite obviously, sex pays its own way even on the basis of short-term benefits.*

The elimination of bad genes through sexual recombination can be of immediate selective advantage.* The Russian A. S. Kondrashov and others have shown that if the effects of deleterious mutations reinforce each other, in the sense that a double dose of defects is more than twice as bad, then the selective advantage of having one or two crossover points per chromosome is quite pronounced. This seems to be in fairly good agreement with what is actually observed. In contrast to Muller's ratchet, this argument does not rely upon the effects of random fluctuations in small populations. It provides an advantage for the individual, rather than the group. But it is a tricky task to check whether the rates of mutation and recombination are of the right size, and above all to verify whether the selection against mutants, in nature, is as intensive as required in the model.

The lure of variety

The most striking immediate effect of sex is surely the variability of offspring. Asexual individuals have (except for a few mutations) the same genome as their parent. The fact that this parent has managed to propagate shows that it is adapted to its environment. If conditions do not change, it must seem the safest policy to stick to the parent's way: *Mother knows best*. Sexual replication, on the other hand, always implies a drastic break-up and reshuffling of the genome. If the number of offspring is large, this may allow survival of individuals over a wide range of possible environmental conditions. The correlation with the parental type will be less pronounced. So if a new type is asked for, it is much more likely to be found among sexually produced offspring than among asexually cloned copies.

The American biologist G. C. Williams used a striking metaphor for this advantage of sex in an uncertain environment.* If you participate in a lottery, he says, you can increase your chances of winning by buying 100 tickets—except if they all carry the same number. 'Multiple copies of the same genome would be as wasteful as multiple purchases of the same lottery ticket.'

Let us imagine, still following Williams, some promising location for a tree in an elm-wood, say. Many elms will cover the patch with their pollen. Thousands of seedlings will start on a fierce race for survival. Only one is going to win and to develop into a fully grown elm. The competition is terrific, and success depends on countless imponderabilities concerning the soil, the weather, the light, and the like. Asexually produced seedlings from an elm will all enter the race with the same genetic program—always the same number in the raffle. It is obviously much better to offer a wide variety of programs for nature to choose from (and so, sure enough, elms do reproduce sexually). The competition among offspring from the same parent, a kind of rivalry between sibs, is counter-productive, from the parents' point of view, if it cannot increase its chances of winning; and such an increase occurs only if the offspring diversify.

This argument from *sib competition* attributes a decisive role to selective advantage. A single number is chosen out of, say, all possible six-digit numbers. It corresponds to the genetic program most closely adapted to the current environment. No consolation prize for being second best. This scenario requires intense competition among numerous offspring. But there are many species which are firmly committed to sex, although sib competition plays no role in their lives—either because the siblings are few, or because they grow up in widely dispersed locations. As Maynard Smith remarked, if one buys a ticket in 100 different raffles, there is no harm if the tickets all have the same number.*

There are several ways of setting up a lottery, by the way. In Williams' example, one number is chosen (by selection), and whoever bought this number wins. In Chapter 4, the winner of the lottery was determined by a different rule. All tickets which had been bought at the raffle were thrown into an urn, and one was then sampled at random. In this case, the number is irrelevant (except for identifying the owner), and whoever purchased most tickets has the best chance of winning. The asexual parent, in this metaphor, is improving its odds by buying one ticket and photocopying it. This model is appropriate if sheer luck, rather than adaptation to local conditions, decides which elm tree is going to grow up. Chance, in this case, weighs more heavily than selection. In such a set-up, sib competition cannot suffice for explaining sex.

The role of sex in 'hedging bets' by diversification has been stressed in many other arguments. One of them has been called the 'Tangled Bank' hypothesis by Graham Bell.* It argues from a point of view which is ecological (or even economical) rather than genetical. The basic tenet is that it pays producers to diversify in a *saturated* economy. As long as soft drink, say, is in great demand, a company's best policy would be to produce just one kind of it, to maximize capacity. Once there is enough on the market, it should diversify its production in flavour, colour, price, etc. to cater for the customers' whims. Similarly, in an environment which is saturated, like the *tangled bank of a river*, for instance (the phrase is from Darwin*), diversification by sex allows more niches to be occupied simultaneously.

This idea is confirmed by the geographical distribution of sexual and asexual species. Thus asexual reproduction occurs more frequently in freshwater lakes than in oceans, for instance. It is found more often at high latitudes, or in recently glaciated regions, while sexual replication predominates in the stable, richly saturated jungles. This not only confirms the widespread prejudice that sex comes easier in the tropics than in cold climates; it also shows that, despite former views to the contrary, sex is *not* more common in uncertain environments. The fact that seasonal switches to parthenogenesis seem mostly to occur when the environment is undersaturated also speaks strongly for the Tangled Bank.*

Escapism

The Tangled Bank hypothesis sees sex as an adaptation to environmental diversity, but stresses *spatial* rather than *temporal* variation. The complementary point of view goes by the name of *Red Queen* hypothesis. The Red Queen, as Lewis Carroll's heroine Alice found out in her adventures *Through the Looking Glass*, lived in a land where it took people all the running they could do to keep in the same place. Whoever wanted to go somewhere else had to run at least twice as fast as that.

Leigh van Valen used this to describe the permanent treadmill of evolutionary adaptation.* Almost any adaptive success of one species constitutes a deterioration of the environment of their rivals, their exploiters, and their prey. Every population is kept on its toes by others who are improving. If it manages to find an adequate countermove, it will only cause the others to try harder. There is no moratorium to stop such coevolutionary arms races. They last for millions of years.*

A sceptic may argue that such environmental changes seem to be far less pervasive than sex actually is. There are many species of deep-sea fish, for instance, which carry on sex since time immemorial, although hardly anything appears to change in their living conditions: always the same low temperature, high pressure, and dark night, always the same silent rain of detritus from upper layers. But what do we know? Environment is not just surroundings. It is also the insides of the deep-sea fish, which are infested by a rich variety of microbes, and this is where things can change very fast.

It was J. B. S. Haldane, it seems, who first pointed out the role of microbial diseases as agents of selection.* Their evolutionary impact was neglected for a long time, possibly because we humans are among the largest animals on Earth and tend to overlook most of the other inhabitants. Factors like climate or the frequency of predators all form part of the environment, and tend to fluctuate over time—but this is as nothing, compared to the speed with which diseases can spread.

If the problem of survival were always asked in the same terms, sexual recombination would seem of little help, since it rakes through the chromosomes as roughly as a poker, blindly destroying tried and proved solutions. In an uncertain environment, change makes more sense. But computer simulations show, somewhat surprisingly, that unstable conditions can offset the twofold cost of sex only if the requirements for consecutive generations are, not just different, but actually almost opposite, or at least negatively correlated.* Such environments are not only jumpy, but downright *contrary* (in Hamilton's expression), perversely bent on defeating whatever proved best in previous generations.

Surely our world cannot be so pig-headed. Or can it? It is here that the parasites, which were press-ganged earlier in the chapter to account for the peacock's tail, display their power to the full. Every higher organism is persecuted by viruses and microbial diseases. Its genetic array of immune defences has to be constantly updated. The immune system acts as a combination lock which the parasites try to break. By trial and error, they always succeed; indeed, they often succeed very soon, owing to their high turnover rate: their generation spans are usually much shorter than their host's. This means that they are always most efficient in attacking the most common immunotypes of their hosts. For the hosts, it can be deadly to adopt a combination code which is currently widespread in the population. It pays to belong to a minority. But if it pays well, in terms of reproduction, this will yield a new majority. Hence each solution is self-defeating in the long run. The host is under constant pressure to change the digits of its combination lock. The best way of doing this is to reshuffle genes, which means by having sex.

The leading (but by no means only) proponent of this theory is W. D. Hamilton, who describes himself as a coevolution freak.* Like G. C.

Williams, Hamilton uses a game to highlight his theory. It is the humble *penny matching* game.* If the pennies match (in our case, the gene arrays of host and parasite), then the parasite wins. This penny matching game seems a rather simple pastime, but it gets interesting if it is played repeatedly. Both players must avoid being figured out by the opponent. They have to randomize incessantly.

Hamilton has more to offer than a metaphor: he has wondrous computer simulations to argue his case* (Fig. 6.6). They show a striking advantage for sex, amply sufficient for offsetting its twofold cost, especially if a large array of genes is involved. The frequencies of these genes are in ceaseless, chaotic

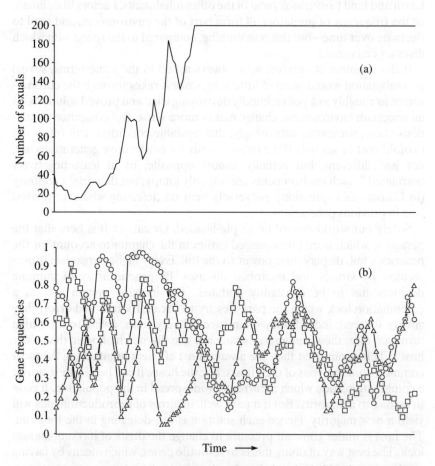

Fig. 6.6 A host population beset by parasites. Population size is 200. Curve (a) shows how sexuals win out. Curve (b) traces the gene frequencies of three genes from the immune system of the host. (After Hamilton *et al.* 1990.)

turmoil: they increase and decrease, but are protected from extinction by constant reshuffling through sex. Genes happening to be out of favour can tide themselves over by boarding offspring together with genes currently in demand. This way of forming short-term alliances makes genes less vulnerable to the vagaries of selection. They acquire enough inertia to carry on through hard times. What is a bad gene today may be a good gene tomorrow, so that it pays to keep them *in circulation*. On the computer screen, parasite and host sweep giddily back and forth, like a couple of waltzing mice (Fig. 6.7). When the genomes are large, the mathematics become intractable, but the simulations speak up quite convincingly.

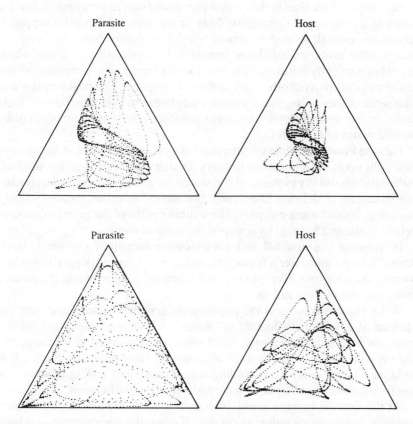

Fig. 6.7 The dance of gene frequencies in a host chased by a parasite. Host and parasite each have three possible genetic alternatives. The better they match, the better for the parasite. The frequencies of these combinations correspond to points on the triangular dance-floor; note how the dancers avoid bumping into the walls (that is, the states where some gene reaches extinction). One gets a vivid impression of their dizzying chase. (After Segers and Hamilton 1987.)

A recent experimental find has brought striking confirmation. It had been known for some time that individual mice can tell each other apart by the odour of their urine, and actually recognize their immunotypes, or more precisely, their major histocompatibility complexes (which seem easier to smell than to spell). It now turns out that this odour is a basis for sexual selection. Female mice wrinkle their nose if the suitor's smell is too much like their own.* They clearly prefer mates with a different immunotype. As a result, this maintains a wide diversity of mechanisms for resisting infectious diseases. Hamilton can hardly have hoped for a more 'pungent' proof.

Another unlooked-for support was provided by Tierra, the artificial ecosystem created by T. S. Ray where self-replicating programs in machine code competed for time in the central processing unit of a computer, and for space in its memory.* Occasional flaws in the copies provided a supply of mutations, enough to feed an almost feverish evolution. Ray's first surprise was that this evolution led almost immediately to 'parasitic' programs which got rid of part of their instructions but brazenly managed to commandeer the corresponding parts of other, self-sufficient programs. The next surprise was that some of these programs spontaneously began to exchange parts of their coded instructions: the cyber-creatures had discovered some sort of sex (they needed mates to replicate).

Parasite Power does more than make sex a plausible activity. It also throws new light upon the shop-worn imagery illustrating the struggle for survival. Adaptation is usually pictured as the search for an optimal solution, consisting ultimately of delicate fine-tuning. But natural selection means not only *adapting*, it also means *escaping*. The writhing paths of the gene frequencies reflect the drama brought by sex upon the stage of nature.

Is evolution like an uphill walk towards some summit of a mountain landscape? It looks more like a frantic attempt to avoid getting bogged down in a swamp. As soon as a hiker slows down when crossing a marsh, the ground gives way and the feet sink in.

W. D. Hamilton pictures the populations' genomes as scattered over the surface of a light ball floating on water.* As soon as too much of the genomes accumulates in a spot, it will submerge and dissolve: for the genes, the only means of escape is swift *dispersal* by recombination. Such a ball bobbing on the water keeps rolling without end (but rolling nowhere), a multi-dimensional analogue of the treadmill of the Red Queen.

At the time of Lewis Carroll, biologists tried to explain sex by its purportedly 'rejuvenating' effect upon cells. Today, this seems naïve, but faint echoes survive in the two main contending theories currently under discussion. One of them stresses, as we have seen, the *eugenic* aspect of sexual recombination: harmful mutations can be efficiently cleared. The other puts the *hygienic* aspect forward: sex helps to fight parasitic diseases. This is not meant, of course, to vindicate the crude slogan that 'Sex is good for your health' which is found on some T-shirts (invariably worn by men). Sex can, of course, lead to diseases. But diseases, apparently, lead to sex. To quote W. D. Hamilton again: 'The theory makes me slightly more at ease about the continuation of my own kind within the human species. I can now tell people that we males are necessary for health.'*

A female which multiplies asexually is a sitting duck for exploiters from parasitic species. Her best countermove is to keep on the move, to stay a moving target by sticking to sex. As it happens, she may then be exploited by the males of her own species. Some feel that this is a switch from bad to worse. But it certainly makes things go round.

7

Evolutionary game theory

Optimal foraging and the ideal free distribution

A battle of the sexes

The theory of games

Simplified poker

Security levels and minimax thinking for zero-sum games

Escalation and the game of 'Chicken'

An explanation for ritual fighting

Payoff and strategy in the biological context

Playing the field versus pairwise interactions

Evolutionarily stable strategies

Fictitious play and game dynamics

Mixed strategies and mixed populations

Symmetric and asymmetric games

Conditional strategies and ownership cues

Population genetics and the tramway effect

Environmental cues and learning rules

7

Evolutionary game theory
Playing for keeps

*. . . Life has only one true attraction, which is the attraction of
the game; but only if we do not care whether we lose or win.*
Charles Baudelaire

Flea waltz

Animals are not good at computing. *Clever Hans*, a horse widely known at
the beginning of the century for doing simple sums, turned out to be cued by
involuntary moves of the questioner.* Some pigeon prodigies can be taught
to count up to eight, and behaviourists are understandably proud of them, but
a normal bird does not count up to three. This is borne out by a trick used by
wildlife filmmakers. Three members of the film team erect a tent in front of
the birds' nest, and enter; then, two of them exit and return to camp, leaving
the third with a camera in the tent. The birds, having anxiously watched the
proceedings, quickly resume whatever they were doing, quite obviously sure
that all is clear.

Fish are no better than birds in this respect.* But in spite of having a weak
head for numbers, they can divide six by three without any coaching, as was
shown by Manfred Milinski. His favourite fish are stickleback. If a lone
stickleback is fed water-fleas from both ends of a tank, it stays at the end with
the higher input rate, not very surprisingly. But if six stickleback are jointly
served the food, their problem is a bit more complicated.* The best solution
now depends on what the others do. It would be silly if all were to queue up at
one end.

In Milinski's set-up, the input rate of water-fleas was twice as high at one
end of the tank than at the other. After a few minutes, the fish settled down to
a pattern: two at the end with the lower rate of supply and four at the other
end. This is a *statistical* distribution: from time to time, a stickleback switches
sides to update its information on water-flea supply, which in nature is not
likely to remain constant. In fact, whenever Milinski exchanged the feeding
rates between the two ends during a run of his experiment, the fish regrouped
within a few minutes accordingly (Fig. 7.1).

Of course, the stickleback do not need to keep tally of their numbers. All

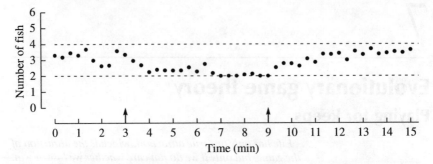

Fig. 7.1 The number of stickleback in the left half of the tank (an average from eleven runs). During the first phase of feeding, which starts after 3 minutes, the supply of water-fleas is twice as high at the right end of the tank; during the second phase, which starts 6 minutes later, it is twice as high at the left end. (After Milinski 1979.)

they have to do is to compare their own food intakes at both ends. In the 2:4 distribution, each stickleback gets one-sixth of the total. A stickleback switching to the other end gets less. Even if it does not 'understand' that it pays to go back, its new neighbours will notice that food has become scarcer, and try their luck at the opposite end. It is much like the molecules of a gas distributing themselves in a container: statistically, they balance out. The fish do not have to be very far-sighted to adopt the *ideal free distribution* (which matches the water-flea distribution).

To behave like molecules is no great feat for stickleback. Their program is more subtle, in fact. When Milinski repeated the experiment of feeding a single stickleback, he noticed that if water-fleas are provided in dense swarms, the fish prefers the *less* crowded regions. This was against all expectation, but it becomes understandable if we observe how difficult it is to concentrate upon one of several moving targets: just try and capture one out of a handful of coins simultaneously tossed at you. If a crowd of prey mills around, a predator is apt to become thoroughly clumsy. The stickleback attempts to avoid such confusion.

But this is not the whole story. Indeed, if the stickleback is *very* hungry, it overcomes the effect of confusion and smartly darts into the densest swarm of water-fleas. It is only when partially satiated that it turns to the less crowded swarm. Milinski was actually able to show that the capture rate of a hungry fish is highest where prey is dense, and that of a well-fed fish where it is less dense.* So the stickleback knows how to choose the region where its food intake is optimal.

But why is a well-fed fish *not* willing to overcome its confusion? It can hardly be laziness. There must be a more respectable reason, having to do with reproductive fitness, preferably. And this reason exists: it is the predator. The stickleback is not just an intrepid hunter of water-fleas. It is hunted in turn. Under normal circumstances, it must always watch over its shoulder and be ready to break off any feasting on water-fleas at the hint of a threat: to flee or not to flee, that is the question. It is only when the risk of starvation is high that it pays to reduce vigilance and to concentrate entirely on the prey.

M. Milinski and R. Heller checked this with an elegant trick.* They flew a kingfisher over the tank (a dummy, of course). As predicted, even the hungriest fish now preferred the *less dense* water-flea swarms. It was obviously too risky to turn a blind eye to the predator. But just as a fighter-pilot watching his rear can give less attention to the front-gun visor, so a jittery stickleback can handle fewer prey at a time.

This stickleback program is already fairly intricate: when hunting, watch your back; if very hungry, throw caution to the wind; but if you have seen a kingfisher lately, don't.

A fish's behaviour can get a great deal more complex yet. In the next example, predators still hover in sight; but hunger is now replaced by lust, and the stickleback by the guppy, who is as bright as the stickleback and no longer than a thumb nail. Anne Magurran and Martin Nowak have found some intriguing patterns in the guppies' version of the *battle of the sexes.**

Sneaky courtiers

Male guppies display an ardour which is truly stunning. They spend almost all their time courting. On average, they try once or twice *a minute* to copulate. Females, predictably, are less enthusiastic. Usually, they are not receptive at all and try hard to escape male attentions.

There are two tactics for the love-struck male. He can display, and hope to be chosen; or he can sneak in and attempt a brief thrust. Any male guppy uses both tactics, with varying frequencies. The display is an elaborate affair, with a spectacular arching of the body and a flurry of the fins. This indicates that sexual selection is very effective. Obviously, a female guppy can afford to be choosy—or rather, she cannot afford not to be: if she lowers her standards, her male offspring are likely to have less-than-average sex-appeal. Therefore a female must spend much of her time avoiding unwanted intercourse.

Now a predator joins the game—a pike-cichlid, for instance, henceforth called a pike. The predator's effect on courtship behaviour is likely to be even more pronounced than its impact on foraging. 'Foraging animals rarely try to draw attention to themselves', as Magurran and Nowak write. 'In contrast, courtship involves elaborate displays which seem as likely to catch the eye of a potential predator as a potential mate.'

The guppies' response to their dilemma can be described in three scenarios. First, if the danger from the pike is very great, it is surely best to concentrate on avoiding it. But having a pike in the tank is not always an immediate threat. Just as grazing antelopes and lions in the Serengeti live side by side most of the time without so much as glancing at each other, so predatory fish and their prey can coexist peacefully in the river or in the fish-tank. Between meals, a pike is harmless, and the guppies can relax their vigilance. Interestingly, male guppies seem more ready to do so than females. In fact, females have cause to be more careful. They are attacked more often, probably for being larger; their countermeasures are effective, however, and they are killed *less* frequently.*

Are female guppies brainier than males? Not necessarily so. A male can have more reason for recklessness. For the male, ignoring the pike means he can resume courtship; for the female, it means she can resume fighting off the males. Apparently, a male has more to gain by achieving a copulation than a female by avoiding it.

So the second scenario—intermediate danger—has the males sex-oriented, but the females still predator-oriented. The females, for instance, inspect the predator by closing in carefully. Males are likely to join these forays, and thereby lessen the risk, but they do not do so, alas, in a chivalrous spirit. In fact, they hang back by an inch or two, and attempt sneaky matings while the female's attention is riveted on the pike. The female cannot do much about it.

There is a third scenario, when the predator's threat is weaker yet. In this case, females can stop watching the pike and concentrate on avoiding importunate males instead. The males, then, have lost their golden opportunity for a sneaky in-and-out; they will do best to forget about it and watch the lurking predator, whose threat, while small, is still present. But then, the females are no longer pestered by the males, and will *also* be able to attend more closely to the pike. Once this happens, the males will be tempted to resume their mating attempts. Which means that females should forget about the pike and deal with the obnoxious males instead. The males, then, would do better to reform—but wait: we have been there before. Our argument has turned full circle, and led us back to square one (Fig. 7.2).

Richard Dawkins introduced a similar 'battle of the sexes' in his discussion

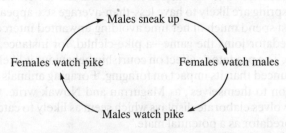

Fig. 7.2 The guppies' battle of the sexes, if the danger from predation is low.

on parental investment.* He described a hypothetical bird species where males were strongly tempted to desert, leaving the task of raising the brood to their mate, and starting a new family somewhere else. The female counter-strategy is to be coy: if the females were to insist on a long engagement, the males would find it too late in the season to start all over again. They should prefer to stay faithfully at home and help with the offspring. If males were faithful, however, females would fare better by skipping the costly engage-ment period. But then, males could again afford to desert; in which case, females would do better to be coy; in which case, males should be faithful, etc. This has the same cyclic structure as the guppies' love-life. Can one step off this roundabout?

Gameboards and battlefields

The mathematics of conflicting interests owes its beginnings to John von Neumann, and goes by the name of *game theory*. Like probability theory, it originated in actual games: but this time, the inspiration came from poker or chess, rather than from dice or roulette. Chance may still come into play, but the main uncertainty arises from the decisions of the coplayers.

Just as the theory of probability eventually found respectable applications in the insurance business, quality control, statistical mechanics, and paternity suits, so the theory of games quickly got rid of its frivolous pedigree and turned to earnest applications in economics and social affairs. It came of age in the 1940s, with the publication of John von Neumann's and Oscar Morgenstern's treatise on *The theory of games and economic behaviour.** During the Second World War, Operations Research had become a major enterprise. For the first time in history, mathematicians contributed decisively to the military effort. The fact that it was all very 'hush-hush' did not do any harm either. Game theory found enthusiastic support.

With the Cold War, many budding game theorists inevitably felt impelled to offer advice to the strategic planners in East and West. They soon got sinecures in government agencies. Research projects promised applications to the tactics of dogfights, the planning of bombing patterns, and the subsequent conduct of peace negotiations; in a flurry of hype, newspapers hailed the new 'science of settling conflicts', and grants started flowing.

These heady days could not last, of course, and in the 1960s, it almost looked as if a policy of overselling had done discredit to the theory of games. Happily, it recovered; but its relations with the military have cooled down. Waterloo may well have been won on the playing fields of Eton, but the theory of games is unlikely to decide wars. It does not even greatly assist in war games and the *Sandkastenspiele* of Staff Schools.

As a matter of fact, it does not help much with chess either, that classical abstraction of military manoeuvring. Game theory offers a proof that there exist optimal strategies for both Black and White—a statement which few non-mathematicians would have doubted at the outset—but it offers no clue as to what such strategies look like. At present, it is unknown whether White's best policy will always lead to a win.

Game theory is not good at prescribing solutions. It does not even tell how to play poker (John von Neumann's favourite game). But it helps in understanding some basic features of conflicts by means of models. One model of poker,* for instance, reduces to two players only, and a deck consisting of two cards, 'Ace' and 'King'. Each player puts a dollar in the pot. Then, the first player draws a card at random and, after taking a look at it, has the option of giving in and losing the pot, or of raising the stakes by adding one dollar to it. In this latter case, the second player, who has not seen the card, can either give in, or else call the bluff by adding one dollar too. The card is then shown: if it turns out to be 'Ace', the first player wins the pot; if it is 'King', the second player does. This game has, admittedly, not quite the thrill of a real poker game, but it displays in a foreshortened form its two main ingredients, bluffing and calling, and we shall have to address it more closely in a moment.

Another textbook favourite has two duellists walking toward each other, each armed with a loaded gun. They can shoot whenever they like—but only once; and if they miss, they are not supposed to run away. Firing too early can be fatal, if one's aim is poor. Waiting for too long can be fatal too, if the adversary's aim is good. Assuming that the duellists know each other's marksmanship, when should they fire? This one is fairly straightforward to solve, but not likely to help you in a real-life duel, unless you can make an inspired guess about the odds of your opponent scoring at twenty paces.

There is a simpler version of the duel, where this sort of inside information is not needed. The so-called 'Chicken' game used to be popular with American teenagers in the days when rock and roll was young. It is very simple. The two contestants race toward each other, preferably in stolen cars. Whoever swerves is branded as a 'chicken' and has lost. This game seems to have limited entertainment value, but it leads, as we shall presently see, straight into biology.*

Examples like 'Chicken' or Russian roulette show that a game can be a matter of life and death. When mathematicians requisitioned the word 'game' to model the most serious conflicts of interest, they did not unduly stretch its meaning. Antagonisms lurk behind most games. A smooth transition leads from lighthearted gambolling to mortal combat;* duels in grand style, with their host of elaborate conventions about armaments, place and time of encounter, the role of seconders, etc. are only gradually different from tournaments, prize fights, sports matches, histrionics, and rites, and in a way closer to the 'spirit of the game' than a flurry of unruly kicking at a children's party would be.

Poker without tears

The two main concepts of game theory are payoff and strategy. A *strategy* is a program for the player, a sequence of decisions. In the simplified version of poker, the first player has only to decide whether to bluff or not; and the second player, whether to call or not. This is the simplest type of game: each player has only two strategies.* Chess is tremendously more complex. A chess player's strategy must in principle specify each move, and therefore have an answer to each possible countermove. In real life, no chess player, whether human or electronic, can possibly be provided with a list of all alternatives. This list is not infinite (because one chess rule says that no position may be repeated three times), but it comes close to it: there are 20 possibilities for the adversary's first move, some 400 positions after the next, etc. So the game theorist's definition of a strategy is an idealization: it compares to the real thing like a geometer's straight line to a stroke of chalk.

When the game is over, each player receives a *payoff*. It need not be money: it can be prestige or fun, a kiss or an injury. Again, the game theorist has to abstract from countless intangible aspects. What matters is that the players should be able to rank the different outcomes by order of preference; usually, by assigning numbers to them.

Here, then, is a game in its mathematical nutshell: each player opts for a strategy and receives a payoff. This payoff depends on the decisions of the players, and possibly on luck.

Take simplified poker, for instance. For the first player, who will certainly not fold after drawing an 'Ace', the strategy *Bluff* means: raise the stakes when you have a 'king'. For the second player, who cannot call if the other folds, the strategy *Call* means: raise the stakes if the first player does. So, if the first player opts for bluffing and the second for calling, two outcomes are equally likely.

1. The first player has an 'Ace', and receives two dollars.
2. The first player has a 'King', and loses two dollars.

The average payoff is zero, in this case. If the first player opts for bluffing and the second for giving in, the first player is sure to earn one dollar; and so on (see Fig. 7.3). In this game the second player is on the losing side, on average; we may imagine that the participants alternate their roles to make up for this.

In contrast to chess, where both players are fully aware of their exact positions, the second poker player knows less than the first. Poker is a game of *incomplete information*. The payoffs are mean values and have to be interpreted statistically. In a long run of poker games, bluffing earns zero against calling, etc. So we are thinking of repeating the game a great number of times—a theme which will lead us far.

	If the second player folds	If the second player calls
If the first player bluffs	1 dollar	0 dollar
If the first player folds	0 dollar	50 cents

Fig. 7.3 Simplified poker. The gains of the first player are shown. Since this is a zero-sum game, they have to be paid by the second player.

If the second player makes a habit of calling, the first player will give up bluffing after a while. By raising the stakes only when holding an 'Ace', one can make sure of an average payoff of half a dollar against a caller. But sooner or later, the second player will notice that the first gave up bluffing, and will stop calling, of course. The first player, who then earns nothing on average, will be sorely tempted to resume bluffing. But then, obviously, the second player will start calling again. We find ourselves back to square one, just as the guppies did.

We need no practice at poker to see that we should sometimes bluff and sometimes not. But when? Should we bluff when the other player has had a streak of losses, and has cause to be in subdued spirits? But the other player could out-guess us. Should we bluff when the other player has a cocky look? But what if the coplayer still penetrates our thoughts? The safest way to mix our strategies is one which *we ourselves* cannot predict: by tossing a coin, for instance.

A program using a randomized decision is called a *mixed* strategy. The second player could, for instance, decide to 'call' each time a Head shows up. It is true that the first player could still *guess* that on average, every second bluff would be called, and figure out that bluffing now yields an expected return of one-half, and not bluffing of one-quarter of a dollar. The first player will then be sure to bluff. The second player will now lose an average of 50 cents per game, and find it better to call more often. Actually, the best strategy for the second player will be to call two-thirds of the time. Then, bluff or no bluff, the first player cannot go away with more than one-third of a dollar (in the long run). The second player has found a way of *minimizing losses*. Similarly, the first player can obtain a guaranteed average income of one-third of a dollar per game, no matter what the other player does, by bluffing with probability one-third (but not, of course, every third time).

Mixing and minimaxing

The simple poker exercise can be extended to a classic result of game theory.* By judiciously mixing their strategies, players can always maximize their minimal payoff, or, what amounts to the same, they can minimize their opponent's maximal payoff. This holds for all *zero-sum* games, where the gain of one player is the other player's loss. Most of the well-known parlour games are zero-sum: the players are at loggerheads, at least as far as payoff is concerned. (They may find common interests in other aspects, like enjoying the fun of the game.)

There is a whiff of pessimism behind such *minimax* thinking in terms of security levels. By assuming that the coplayer is going to find the most hurtful reply, minimax evaluates the strategies according to their worst possible outcome. This is, incidentally, the official doctrine of decision in the US Armed Forces: to base strategic choices not on the enemy's intentions, but on the enemy's capabilities—on what the adversary is *able* (rather than likely) to do.*

The German General Staff held the other doctrine. The fact that it lost out may have been due to other reasons. But the conservative approach of reckoning with the worst obviously has much in its favour. Above all, it credits the adversary with an at least equal amount of intelligence. Countless defeats are due to *downplaying* the adversary's threat.

However, the minimax philosophy works only if the other player is really out to get you. If your win is the other's loss, this is a reasonable assumption. But there are many situations where the interests are not totally opposed. Even in war, the two parties are likely *both* to wish to avoid certain outcomes. Many game theorists nowadays feel that the early emphasis on zero-sum was overdone. It led to some remarkable results, but underrated other types of interesting conflicts.

Take 'Chicken', for instance. While there is plenty of antagonism in this game, the two drivers are not entirely at odds: both agree on *not* wishing to crash. Where they disagree is on who should do the avoiding. In a real 'Chicken' game, the payoff is exceedingly complex, involving repair costs, the danger of being caught by the police and, of course, 'loss of face' in more senses than one. Again, we turn to a simplified model* with the following rules: if both drivers swerve, no harm is done; if no driver swerves, each has to pay 100 dollars for damaging the cars; and a driver swerving unilaterally must pay 10 dollars to the driver who kept on a straight course (Fig. 7.4).

How would *you* play this game? Your worst outcome—the crash—is worst for your adversary too. By swerving, you maximize your minimal payoff. By not swerving, you minimize your opponent's maximal payoff—no longer the same thing. You are in the same position as your adversary, but you should try to do whatever the other *does not*. Unfortunately, you do not know what

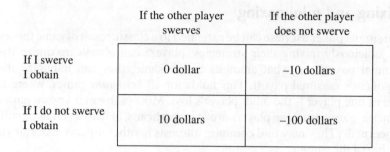

	If the other player swerves	If the other player does not swerve
If I swerve I obtain	0 dollar	−10 dollars
If I do not swerve I obtain	10 dollars	−100 dollars

Fig. 7.4 The game of 'Chicken'.

the other will do. You can try to guess, of course. It is actually easy to see that if the odds that your adversary will swerve are higher than 90 per cent, it is worth taking the risk of steering straight; if the odds are less, you should swerve. If the odds are precisely 90 per cent, your two strategies will lead to the same average payoff, namely the loss of one dollar.

But does this give you a clue to your opponent's most likely decision? You can, of course, reckon that the other player will also be able to do sums, and would drive straight if your probability of swerving were known to be more than 90 per cent; it would be better for you, then, to definitely swerve. But if your probability of swerving were known to be less, the other would swerve; you would do better, in that case, to drive straight. Is there any reason for you to swerve with exactly 90 per cent probability? The adversary would no longer know what to do, but so what?

It is true that the 90 per cent solution has a kind of inner consistency. If your adversary swerves with a probability of 0, or 1, or 2 per cent, etc. up to 89 per cent, you should swerve, while for 91 per cent up to 100 per cent you should not. If all these percentages were equally likely, you should swerve for 90 per cent of them. In this case, your opponent has no reason to swerve with one probability rather than another. This may seem to imply that each percentage should be equally likely, and so to wrap it up. But if you think so, then put yourself in your adversary's shoes. If you could swerve with any probability you like, why should you swerve with 90 per cent? After all, you play 'Chicken' for dollars and not for consistency. Most game theorists agree on 90 per cent as the 'rational' solution, but the argument is somewhat tenuous.

The rational reader, of course, should not play 'Chicken' at all. Nevertheless, each of us plays similar games many times a week. We often have to decide whether or not to *escalate* a conflict. In most cases, to insist does not pay except if the other gives in. Only lawyers seem invariably to gain from escalating. An ordinary person cannot make a stand for every trifle. 'One war at a time', as Abraham Lincoln said. On the other hand, nothing could be worse than always retreating. A mixed strategy is clearly called for.

Whether to intensify a conflict or not is a daily issue for policy-makers. Their playing with fire usually involves elaborate signalling and probing, 'brinkmanship' and 'flexible response'. But when the trimmings of diplomacy and rhetoric are swept away, the bare bones of 'Chicken' are apt to reappear.

Once a stag, always game

Is escalation worth the risk? This basic question confronts animals as often as statesmen. Although fighting provides some of the most colourful aspects of wildlife, animals have frequently been observed to refrain from escalating contests. Rather than tear at each other with claws and teeth, they may just engage in a staring contest, or a trial of strength. Sometimes, they literally *swerve* away from their collision course, rather than use their horns, tusks, or antlers.

One of the best studied types of conflict is played out between deer stags.* Typically, such a contest starts with a roaring match; this is followed by a characteristic *parallel walk* of the two antagonists; this in turn leads to a head-on clash and a strenuous pushing match with interlocked antlers; and only then, as a last resort, comes the all-out fight which may cause mortal injury. The preliminary bouts serve for mutual assessment; the roar carries information about the body weight, the parallel walk allows the animals to compare sizes, and the pushing match is a fair trial of strength and stamina. Each stage allows the antagonists to gain extra information, but at an increasing risk; each stage can end with the retreat of one of the rivals, but the chance to withdraw with impunity shrinks as the contest continues.

During the parallel walk, it may happen that one of the stags turns prematurely toward his adversary, ready to start the pushing match. The opponent's flank must be a hugely rewarding target for his lowered antlers. Usually, however, the stag withholds his attack and resumes the parallel walk, catching up in a few quick steps with the unsuspecting rival. This restraint has been much commented upon by students of animal behaviour.* It reminds one of the Queensberry rules: boxers have to wear gloves, they must refrain from hitting below the belt, they are to wait in the neutral corner of the ring if their rival is floored, etc. These are clearly civilized rules, and a referee sees that they are obeyed. The stags do it without a referee.

Such *ritual fighting* is obviously good for the species, but evolutionists know that this advantage cannot explain it. A stag deer killing his rival at the first opportunity, without bothering with the conventional preliminaries of escalation, would earn himself a large harem and multiply much faster than average. He is not likely to be inhibited by the 'thought' that if all stags were to adopt his recipe, they would decimate their population. A trait with an immediate advantage to its bearer will spread, even if this dooms the population. And the trait of killing a rival stag without any ado has been observed to exist, if only rarely. What prevents its spreading? The gamekeepers, nowadays. They shoot the outlaws. But what about earlier times, and other animals? Ritual fighting has been observed in countless species.

The first to explain it in Darwinian terms were J. Maynard Smith and G. A. Price.* Essentially, they used the 'Chicken' game to model the conflict. We have seen how difficult it is to justify why in a single such encounter, a player should swerve with 90 per cent probability. But the two players are now supposed to be members of a large population and to meet randomly. It makes sense, here, to assume that its members 'swerve' with a probability of 90 per cent. If the odds were higher, escalation would pay, and become more likely; if the odds were lower, swerving would spread. Within a population, *self-regulation* would adjust the average probability of escalation at 10 per cent, because the players now experience the average level of escalation, react to it, and thereby affect it. They become part of the 'others' for all others. Both have been honed against the same population of opponents. As the saying goes, 'We have met the enemy, and he is us'.

The strategy of escalating with 10 per cent probability is *evolutionarily stable*; which means, as Maynard Smith put it, that 'if most members of a large population adopt it, then no mutant strategy can invade the population'.* This requires in the first place that all available strategies fare on average equally well, for if, say, 'escalate' did better than 'retreat' in the 10 per cent mixture, then a pugnacious mutant would multiply, and vice versa. But this *equilibrium* property is not, by itself, enough to guarantee *stability*. Since a mutant ready to escalate fares exactly as well as the rest of the population, it will not be selected against and its copies may still spread by neutral drift. It is *only then* that they begin to be discriminated against. Their increase tilts the balance against their own strategy. In an environment where more than 10 per cent are prepared to escalate, it would be better to swerve. In an encounter with their like, those who escalate the contest fare less well than average.

The net outcome is that only one conflict out of one hundred may lead to injury (for both opponents must be willing to escalate). This is unquestionably 'good for the species', but it would be better still if *all* its members were willing to retreat rather than engage in injurious fights. This cannot be, however, since it would offer ideal conditions for mutants prone to escalate.

It hardly needs saying that the level of 10 per cent follows from our arbitrary choice of payoff values: it is the ratio of possible gain (10 dollars, if the other is a 'Chicken') to possible loss (100 dollars for the mechanic's bill). If losses were higher, the escalation level would be lower still. For heavily armed species like stags or wolves, the risk of injury is fearsome. Such animals stick to ritual fighting and escalate very rarely. Doves, on the other hand, have little to fear from injury (if one dove is getting the upper hand, the other just flies away), and so they escalate freely. If two doves are kept in a small cage, the weaker will slowly be pecked to death.* But doves have been symbols of peace since Biblical times, and journalists during the war in Vietnam used to label as 'Hawks' those willing to escalate and as 'Doves' those that were not. Maynard Smith knew better, but kept the terms. They were intended for a thought experiment only.

It is obvious that for actual populations, many more aspects have to be considered.* The 'Hawk–Dove' model was meant as a starting point. It met with instant success, and the theory of games gained a firm foothold in biology.

Brood and games

Since ancient times, plays and games have been likened to life. My most venerable testimony is a passage from Plato: 'Life must be lived as play, playing certain games.'* But this is unlikely to be the oldest specimen of life/game metaphors. In view of their prevalence, it may seem surprising that it took from World War II to Vietnam for games to be adopted by theoretical biologists. But it needed a new interpretation of the basic notions of *strategy* and *payoff*.

In most human games, social and psychological issues make it difficult to rank the different payoffs. How is one to arrange money, health, self-esteem, or fun on a common scale? Games devised by psychologists have shown in a number of instances that the task may indeed be hopeless: players definitely prefer outcome A to B, B to C, and C to A. Some trace of how diverse the stakes of games can be is shown in the three terms 'price', 'praise' and 'prize', which have a common root and mean very different types of payoff. However, no ambiguity occurs in biological games. Payoff now has nothing to do with preference at all. It simply means increase in fitness. There are, of course, considerable difficulties in determining fitness. But in principle, it is simply the expected number of offspring, and there is no problem in ranking that. As for the strategies, all connotations of plotting and scheming familiar from war or from chess disappear in a biological context. We are now engaging in a 'game theory without rationality'.* It is not conscious cost–benefit analysis that causes an animal to escalate.

It was soon understood that game theoretic models applied not just to fights, but to any conflict of interests—for instance, between male and female fish on the advisability of copulating, or between birds and chicks on when parents should stop feeding the young. Direct confrontations between pairs of antagonists form only a fraction of the struggle for life, incidentally. The success of programs affecting litter size or foraging tactics depends not on the action of single opponents, but rather on the whole population. Mostly, one is *playing the field*.*

Fisher's sex ratio theory, which we know from Chapter 6, offers a good example. The strategies are the different sex ratios. They can produce more or fewer offspring (in the grandchildren's generation), depending on the average sex ratio *in the population*. If this average is male-biased, a female-biased litter has a higher payoff, and vice versa. The 'decisions' here are not conscious, of course; they cannot even properly be ascribed to individuals, since it is *pairs* that produce offspring. In fact, the decisions have been made by natural selection. Their payoff does not depend on what any specific individual or couple of coplayers is doing, but on the total frequencies of daughters and sons.

Fisher's explanation predates game theory, but uses the same kind of thinking in terms of strategies, populations, and equilibria which, in the hands of John Maynard Smith, became such a powerful tool. (In another of his papers, interestingly, Fisher analysed a century-old problem related to an old French card game,* and discovered the advantage of a randomized strategy in what was a special case of John von Neumann's minimax theorem, again without being aware of the more general aspects of the theory of games, which by then was in its embryonic stage.)

An early attempt at using game theoretic arguments in evolutionary biology was made by the American biologist Richard Lewontin, who in 1960 discussed populations playing 'against Nature'.* Their payoff was the survival of the species, somewhat in analogy to the fact that in economic life, firms and corporations are often more interested in ensuring their survival, rather than in maximizing their profit. True to the minimax philosophy, Lewontin was looking for answers against worst-case scenarios, and found that mixed strategies were useful in 'hedging the bets'. But his paper bypassed what later became two main points of evolutionary game theory—the modelling of *local competition* within a population and the fact that a strategy's success depends on its *frequency*.

In 1967, W. D. Hamilton re-appraised Fisher's sex ratio theory, extending it to set-ups where random mating (one of the main, albeit implicit assumptions of Fisher) fails to hold. Hamilton stressed the game-theoretic aspect of Fisher's approach in his concept of an *unbeatable strategy*:* a strategy which, if all members adopt it, offers no scope for individual improvement. This notion and the 'Hawk–Dove' model were the two major paradigms of evolutionary game theory.

Common to both was the implicit assumption of an underlying *dynamics*. The equilibrium value for both the sex ratio and the level of escalation was stable in the sense that selection would counteract any deviation. The simplest candidate for such a game dynamics* is obtained by assuming that a strategy spreads in the population if its payoff is above average; more precisely, that its rate of growth (or decay) corresponds to how much it fares better (or worse) than the mean. Since the payoff is supposed to be measured in number of offspring, this seems natural enough. ·

It is true that a more careful look reveals a snag. Must the offspring inherit the same strategy? Not necessarily. It need not be innate in the first place. Even if it were, what is the part of each parent? But let us shelve these questions for the moment, and treat game dynamics as a convenient *short cut* in evolutionary biology. When Maynard Smith speaks of 'mutant' strategies, this suggests that the behavioural programs use genes to spread. But other ways of spreading are conceivable—spreading by imitation, for instance (like yawning) or by infection (like sneezing). Right now, let us not be particular about it.

Fictitious moves and game dynamics

The dynamic aspect of evolutionary games would have pleased John von Neumann, who expressed some regret at the static nature of his economic theory.* But that shortcoming was partly cured in the early 1950s, when G. W. Brown discovered an ingenious way of finding equilibria in zero-sum games.* This is the method of 'fictitious play'.

Let us suppose that our poker players keep repeating their game, each time choosing the best reply for what their opponent has done so far. More precisely, the first player has to remember the opponent's moves in all the previous rounds, and compute their average; then, the player must select whatever *pure* strategy (bluffing or not) is the best reply to that. The second player, however, is not going to play this average strategy, but the pure strategy (calling or giving in) which is best against the first player's average. Brown proved that this simple recipe leads the two *averages* toward the minimax strategies. To my feeling, it provides one of the best justifications for viewing these strategies as 'solutions' of the game. Besides, this groping toward a solution by *recurrent updating* has a distinct biological touch.

In evolutionary game dynamics, it is not *two* players, but large populations adjusting their strategies. It does not matter whether individuals repeat the game many times, or find themselves confronted with it only once in their lifetime. The evaluation is done within each generation. Each strategy may encounter a variety of replies. What counts is the average, again; it is not an average over the past, however, but an average over the *present* population. And the updating is not an immediate shift to whatever is optimal, but a slow change: strategies currently doing best spread at the other's expense. By becoming more frequent, they may alter their own strategic environment and thereby affect their future success. The most frequent strategies will also spawn the most mutants, who continuously test a variety of alternatives. It is like the march of an army sending out scouts to explore the vicinity and moving into the direction which shows most promise. By moving there, the army alters the conditions found by the scouts, of course.

If there are many strategic alternatives, this kind of adaptive dynamics is liable to become very complex. We may expect the same range of cyclic and chaotic motions as with ecological population dynamics.* But if the game admits only two alternatives, there are just three possible dynamics, as for any two competing populations.

1. One alternative may *dominate* the other, in the sense that it always fares better, no matter whether it encounters the other strategy or a copy of itself. In that case, the whole population will adopt it in the long run. Such a strategy need *not* be good for the population, by the way. For instance, if forest trees have the option of growing an extra metre, they should take it up, no matter whether their neighbours do it or not; for in the latter case, they can over-shadow their neighbours, while in the former case, they can prevent being overshadowed by them. Hence trees will have to spend more and more of their resources on growing. It seems a wasteful folly, but there is no way to reach a consensus to stop it.

2. In a *bi-stable* game, each of the two alternatives is the best reply against itself. One of the alternatives will win, in that case, by reinforcing its initial advantage. The classic example: in Britain, it is best to drive on the left, but not so on the Continent. This example suggests that there is little reason to favour one alternative over the other, but this is misleading. Even if there *were* a reason (if all cars had their steering wheel on the left, for instance), drivers could not simply switch to the right lane in Britain. A population can be trapped by one alternative and constrained to keep to it.

3. The case of *coexistence*, finally. Each strategy is a best reply against the other, but not against itself; therefore, it cannot overtake the entire population, but must live side by side with its alternative. The 'Chicken' game provides an example. Another example is the vigilance game played by foraging animals, for instance juncos.* They have to divide their attention between feeding themselves and scanning for predators. Juncos whose neighbours spend a lot of time scanning can safely concentrate on feeding. But if the neighbours do not look up from their plate either, the danger from predators increases, and your junco should be on guard. Again, a stable balance will evolve.

More generally, if the alternatives become advantageous when rare, they should coexist. In genetics, such a *rare allele effect* was first pointed out by J. B. S. Haldane,* and plays the central role in W. D. Hamilton's theory of sex. It may well be the main prop of nature's variety.

All mixed up

Within a population, pluralistic behaviour can be realized in two ways: (a) all individuals can play the same *mixed* strategy, using one alternative or another with this or that probability; or (b) the population can consist of several types of individuals, each firmly wedded to one of the alternatives. In the vigilance game, for instance, the population could be a mixture, with half of the juncos never scanning, and the others scanning 20 per cent of their time; or else, all juncos could scan 10 per cent of their time. In a game-theoretic analysis of the strategic alternatives, this would lead to the same result. But a biologist will rightly deem the difference important: in the latter case, the population is homogeneous, and in the former case it is split in two types. It could of course also be split into three, eight, or twenty types, each with its own likelihood of scanning.

The idea that animals use mixed strategies has met with ridicule. A soccer player who throws a coin to decide whether to shoot his penalty into the left or the right corner of the goal would be deemed eccentric. And yet, a player using any other way to arrive at his decision must fear that the goalie can figure him out. Similarly, a hare chased by a wolf should show no predilection for swerving right rather than left. Nevertheless, many ethologists are wary of speaking of a 'randomized' decision. They argue that if the panting hare swerves left, this is probably due to a number of factors, like the unevenness of the ground, the angle to the sun, the position of the pursuing wolf, or the flutter of a bird suddenly taking flight. Similarly, the soccer player will be influenced by the last few penalties, the posture of the goalie, the position of the other players, the wind, and a roar from the sidelines. But we know that very complex causation may reduce to sheer randomness. What matters is that no one can anticipate the outcome; not even the player, who could inadvertently give the game away. From this point of view, the superstitious ceremonies used by Roman emperors and generals to reach their decisions make some sense. The fact that the first bird happening to pass by came from the left was understood as a hint from the gods; but in effect, it was used as a random device, and led to mixed strategies without any help from game theorists. Augurs sufficed.

But back to biology: it has been shown that some strains of rats, in a maze, tend to turn left more frequently than right.* Psychologists dub this an 'innate hypothesis': the mixing of alternatives need not be 1:1, but can be *biased*. There is little doubt, incidentally, that if hares had a similar bias, wolves would—quite literally—quickly catch on.

Animal behaviour abounds with examples of mixed strategies, in a broad sense. The foremost example is the sex ratio: usually, it is not 0 (only girls) or 1 (only boys), but some mixture. The mixed strategy of a couple is not the actual sex ratio of its offspring, by the way (since it is a small sample, the frequency of daughters need not correspond to their probability). The game-theoretic argument alone does not allow predictions on *individual* sex ratios. All it can show is that the *overall* sex ratio will move to an equilibrium.

Similarly, the 'Chicken'-model does not predict whether the population will be homogeneous or not. It takes in each case real data to decide the issue. For instance, W. D. Hamilton has pointed out that certain fig wasp populations consist of two types with different fighting behaviour.* Some of the males are wingless, consist mostly of mandibles, and fight it out on the spot, often with lethal results; others have wings, avoid fights by dispersing, and look for places provided with females *not* guarded by mandible-males. There is striking (if indirect) evidence that both types of males achieve equal success at mating: the fraction of dispersing males corresponds to the fraction of unguarded females. Furthermore, such a balance appears to be stable. If more males fight, dispersing gives the better payoff; if more disperse, staying does.

Role-playing

Game dynamics leads to a stable outcome for games with two alternatives. But in other situations, there need be no tendency toward an equilibrium. Evolutionarily stable states need not exist; if they exist, they need not be attainable (as we shall see in the next chapter).

The inadequacy of a static approach is well illustrated by the guppies' battle of the sexes. In the baffling case of low risk from predation, there exists an equilibrium, both for the males and for the females, between their two alternatives. But these equilibrium values are regulated by the *other* sex. This means that if the mixed crowd of males is in the right proportions, females watching the pike have the same average payoff as females watching the males; and that similarly, as long as the females are in the right proportions, males sneaking up for a mating have the same payoff as males keeping an eye on the pike instead. Hence, neither males nor females have any particular cause for adopting another mixture of strategies; but neither have they any cause *not to*.

Suppose, for instance, that some extra males switch from harassing females to pike-watching. They are not penalized for it; they fare neither better nor worse than their sneaky comrades, and there is no selection pressure for them to revert to their old ways. The females, too, do not fare any worse after this change in the males. In fact, they fare better: less harassment. But, and this is the catch, the improvement is greater for females watching the pike than for those watching the males. The danger from sneaky males has decreased and no longer deserves so much attention. Hence, the proportion of pike-watching females will grow. It is only now that males are faced with the results of their shift in behaviour. They will *all* profit from it, by encountering less suspicion from females. But sneaky males have a larger share in this profit, and will spread. This does not mean that the previous shift from sneakers to non-sneakers is simply cancelled, and that everything returns to normal. With so many distracted females around, the proportion of sneakers will *overshoot*; and then the females will respond by reversing their policy, keeping an alert eye on the males, and overshooting too.

It is hard to figure out what is going to happen in such an unstable set-up. If the reaction to altered conditions were instantaneous, the frequencies would periodically grow and fall; with a delay in the reaction, the swings upward and downward will actually increase, and lead to more and more violent oscillations. But if the brains of the guppies keep track of a large portion of the past, and average over their histories, their overshooting would probably dampen down and their strategies tend toward the equilibrium solutions, as with fictitious play.

In contrast to 'Chicken', the battle of the sexes is an asymmetric game: the players have different *roles*. Such situations are very common in biology, and even in the artificial world of 'fair' board games and sporting encounters. At chess, White has a distinct advantage; a football team is favoured on its own field; for the simplified poker game, the first player can expect a higher payoff, and so on. Usually, some rules aim at reducing this asymmetry. A coin is tossed to decide who makes the first move on the board; a match is followed by its return match; poker players alternate at dealing out the cards, etc.

It is obvious that in nature, asymmetries tend to be more pronounced. In games between males and females, parent and offspring, queen and worker bee, or prey and predator, the player's roles are totally different. Even in seemingly symmetric encounters, as for instance between two rivals fighting for a female, the contestants will in general not be as well matched as bantam boxers, say. Much of the ritualized part of an animal fight can be viewed as a test to assess asymmetries concerning size, endurance, and motivation.*

Roles can change during the lifetime of an individual. An intruder can become owner; an offspring, parent; a weakling, champion or vice versa. Worker bees can be promoted to queens. In some species, individuals can even change their sex. It is likely that in many such cases, conditional strategies have evolved, saying, for instance, *if young, insist on a long weaning period; but when an adult, revise your stand*. And a population may seem to exhibit highly variable behaviour, although all its members adopt the same strategy: *if you are weak, retreat; if you are strong, escalate*. This strategy is conditional, but *not* mixed. There is no randomness in such a decision, and nothing balanced in the population. Escalation and retreat yield different payoffs. The weaker individuals are simply making, in Maynard Smith's expression, 'the best of a bad job'.

A general result from the German game theorist Reinhard Selten states that in asymmetric games, there can be no evolutionarily stable strategies which are mixed.* Essentially, this is because a strategy can never encounter copies of itself, as when Hawk meets Hawk in the symmetric 'Chicken' game, for instance. With asymmetry, the feedback is not head-on, but roundabout. A sneaky male does not suffer directly from there being too many of his like around, but only indirectly, from the fact that this causes the females to be watchful.

The simplified poker game is asymmetric: it displays the same tendency to cycle as the battle of the sexes. There is no evolutionarily stable amount of bluffing. In animal contests, a mane can convey false information on size, for instance, or a threatening gesture false information about the intention to attack, but we must expect such signalling to lose its effectiveness after a while. Maynard Smith argues that only surrender signals allow reliable predictions. Bluff and other forms of cheating are transitory.* This does not mean, of course, that they are rare. *Delusions* belong to the game-sphere, as the Latin root *ludus* shows. Cheaters keep the ball rolling, just as other parasites do.

The discreet charm of the bourgeoisie

Possibly the best studied biological game with two roles is that between intruder and owner. As with 'Chicken', neither escalation nor conventional display is evolutionarily stable; but John Maynard Smith and Geoffrey Parker have shown that there is a conditional strategy that fits the bill, namely *escalate if you are owner; if not, do not*. This program uses the asymmetry between the contestants as a cue. It will not pay to deviate from the strategy if all others stick to it. An intruder should retreat, since the opponent is bound to escalate. An owner should make a stand of it, since this means no real risk. If all adopt that policy, which has been termed *Bourgeois* by Maynard Smith, there will be no escalated conflict at all; disputes are settled *as if* by consent.*

It must be stressed that this holds if the value of winning the fight is less than the cost of injury. If a territory is absolutely essential for reproduction, and a vacancy unlikely, then a Bourgeois' proper regard for 'property' makes no sense and should yield to what Alan Grafen calls the *desperado effect*.* Grafen also underlines the fact that the value of winning is higher in a population that respects ownership: territories are more likely to be retained for longer. Paradoxically, this makes the Bourgeois strategy somewhat less plausible.

Nevertheless, many animals use it, as has been verified by fooling them about their roles. If two speckled wood butterflies contest the same sunny spot, the intruder departs after a short spiral flight; but if both butterflies are led to perceive themselves as residents, they prolong their contest, and either one may end up as the winner.* Since a sunny spot never lasts for long (especially in Britain), its value is small, which favours the establishment of a Bourgeois strategy.

There are usually many asymmetries more directly relevant to the outcome of a fight than the ownership cue, foremost being that between weak and strong. This is likely to override other considerations. The odds are that the owner will be the stronger, having won at least once before. (And the fact that the majority of boxing championships end with a defeat of the challenger can likewise be explained without assuming that he was loath to escalate.) On the other hand, experiments on male baboons have shown that previous owner-ship may indeed be decisive: if the roles of the two contestants are reversed, the outcome is reversed too.* Again, if *both* baboons have reason to perceive themselves as owners, the fight escalates viciously.

Humans also tend to adhere to the Bourgeois strategy. It feels eminently reasonable, but may lead to apparently irrational behaviour. Take, for instance, what has been called the *endowment effect* by experimental psychologists.* A person has been given two tickets to the opera and is offered, a few days later, 200 dollars for them; or else, the person has been given 200 dollars and can, a few days later, buy two opera seats for that price. In both cases, most people will refuse the exchange. They seem to prefer the seats to the money, and the money to the seats. In fact, the contradiction is only apparent; what most people prefer is to keep what they *own*—which makes good Bourgeois sense.

What seems to make less sense is that the opposite conditional strategy, *if intruder, escalate; if owner, retreat*, can *also* be evolutionarily stable. If all adopt it, there will be no escalation; but whoever deviates gets embroiled in ceaseless fighting. There is a spider species reputed to follow this freakish strategy: the owners of webs take flight at the sight of intruders, only to take over, apparently without effort, their neighbour's web.* To most of us, this rule appears bizarre; but to a utopian socialist, it must seem perfectly fair. After all, the owner has had occasion to enjoy the property, so now it is the next person's turn. 'Expropriate the expropriators'—a slogan quoted by Karl Renner, first Austrian chancellor after the War, when, on being offered a cigar at Soviet Headquarters, he pocketed the whole box.

Designer genes

'I would hate to commit myself to the view that our bias in favour of Bourgeois is innate rather than acquired', writes Maynard Smith.* So far, we have studiously avoided taking a closer look at how strategies actually spread. But now we should face the issue. In the first place, we must ask whether the transmission proceeds by genetic or cultural means.

It depends, of course. With insects, the behaviour is usually so *hard-wired* that its genetic basis can hardly be doubted. In a few cases, this has been rigorously proved; as, for instance, with the preference of female ladybirds for this or that type of male,* with a bee's program for removing diseased larvae from its cells,* or with a field cricket's propensity for calling to attract mates.* With fish, mammals, or birds, the program is in general far more open. But the cuckoo obligingly provides a particularly convincing example of innate behaviour. The cuckoo female never meets anyone who could show her what to do with her eggs: yet she knows.

'Nature' and 'nurture' work hand in glove, usually. There is some inborn readiness to react to this or that external cue, to learn such and such a sequence of moves at one or another stage of the development. It is difficult to trace specific instincts back to specific genes. But we know that hormonal levels, for instance, are at least partly set by the genome; some such gene constellations may make aggressive behaviour more likely, and can be viewed, therefore, as genes for aggression. In a similar vein, some genes will affect the sense of equilibrium or the level of emotional instability and may therefore be viewed as genes for skiing or for poetry, although they were clearly not selected for these activities. In the same sense, one may assert that there are genes for intelligence, or beauty, or even genes for making human behaviour less hard-wired than that of ants—programs to fuzzify other programs, so to speak. All this has been argued by E. O. Wilson and Richard Dawkins with such success that it is hard to appreciate nowadays how shocking these views seemed to a generation of intellectuals who scented racism behind any talk of human nature.*

For genetically determined strategies, the evolution takes place in the gene pool. Game dynamics must then reduce, in principle, to a branch of population genetics dealing with frequency-dependent selection. The success of an allele for, say, mandibles in male fig wasps depends on how many of its like are around.

In game dynamics, one assumes that the success of a strategy is measured by the number of its copies. But offspring are *not really* copies. 'Like begets like' is, at best, a caricature of the complex Mendelian inheritance mechanism. It is easy, therefore, to throw genetic spanners into the works of game theorists. If an evolutionarily stable program were only played by heterozygotes, for instance, then it could never make up the whole population. This argument, however, is not specifically directed against game theory, but against any adaptationist explanation.* We know that polar bears wear white because this is most suitable for hunting in the Arctic, and we need not worry about what would happen if homozygote bears were zebra-striped.

It has been shown that *usually* the strategic analysis points the same way as the genetic evolution. It may happen that the genome is not able to follow the game theoretic pointer all the way, for instance because some of the strategies are just not feasible. This is clearly a major headache for the model builder:

how can one circumscribe the range of genetically plausible programs? A device to breathe fire might do wonders, for instance (at least as long as it is rare and not used in dry woods), but it cannot be realized. Within reasonable bounds, however, the genetic implementation obeys the strategic advice, at least by doing the next best thing possible. Population geneticists have shown for the sex ratio and other examples that if a new mutant can improve its strategic standing, it will spread.* And Peter Hammerstein's *streetcar theory* shows that in genetic equilibrium, an evolutionarily stable strategy acts as a final stop—no new gene can lead the population away from it. In general, of course, we cannot set up our analysis on the level of allele frequencies, as we know painfully little about the pathways from genes via enzymes to hormones and muscles and nerves. Actually, even if we knew, we would end up with a thicket of impenetrable equations, which is just as painful. The streetcar effect implies, however, that there is no reason to query the broad conclusions of game dynamics: we may trust the genes to play ball, and to assist any strategic design.

Reward and punishment

Most bodily traits develop through an interaction of genetic and environ-mental factors. It is likely to be the same with behavioural patterns. In higher animals, many such patterns are acquired by learning, either through direct experience or through imitation of others. Some ways of transmission are truly stunning: male cowbirds, for instance, can even learn the song preferred by the females of *another* subspecies, in spite of never having any occasion to hear it.* They just go by watching the reactions of their audience.

It seems plausible that the right strategy should be learned (rather than inherited) whenever the payoff is subject to change and has to be evaluated 'on the spot', so to speak. The amount of cooperation or the distribution of food within a society tends to be rather variable, and an individual's response needs frequent readjustment. In these cases, a hard-wired program would fare less well than a *learning rule*. The sex ratio in a human population, on the other hand, is not subject to great fluctuations, so that it should make sense to encode the right strategy once and for all into the genes. If occasional re-evaluation is unnecessary, there is no point in learning rather than trusting a hard-wired program. However, this deserves a cautionary note. A duckling, for instance, *learns* whom to mate with through sexual imprinting at an early stage; if it is raised by foster parents of another species, it will later on look out for a mate of that species.* So here a program is learned although it would appear rigid enough to be committed to genes.

Experimental psychologists have studied learning by confronting pigeons, pigs, or trout with automata resembling those found in Reno or Las Vegas (but less flashy). It turned out that most animal learning is very specific, responding during sensitive periods to precise cues. For instance, it was shown that by punishing rats with X-ray induced sickness, they can be taught to avoid food of a given flavour, but not food of a given pellet size; while by punishing them with electrical shocks, they can be taught to avoid pellets of a given size, but not food of a given flavour.* Sickening as it literally is, this experiment dispels the widely held belief that any stimulus can be used to teach any response.

A learning rule is a program to set up a program, a long-term strategy for choosing between short-term strategies. As such, the learning rule itself can be inherited or learned; we may think of learning rules for learning learning rules, etc. For the sake of an example, let us plunge back into the tank where the six stickleback thrive on their water-fleas. The fish do not need many generations to reach the ideal free distribution which is the optimal solution to their problem. They do it within a few minutes. Somehow, they manage to converge upon the correct decision by a dynamic procedure.* The usual game dynamics, which has strategies spawning off copies in proportion to their success, is only a metaphor. Just as with genetic transmission, it has to give way to a more realistic model in the case of learning; again, there are many possible mechanisms. Again, in the absence of more information, one can but explore some simple and plausible candidates.

Such a candidate mechanism has been proposed by C. B. Harley.* His rule repeatedly adjusts the probability for using this or that alternative according to success so far. Recent experience counts more than past. The eventual outcome remains subject to occasional re-evaluation. For this reason, there must be some fuzziness in the rule which allows it occasionally to try the wrong thing.

Harley applied this rule to Milinski's stickleback experiment and found that it led swiftly to the ideal free distribution. Although in the model all fish obey the same program, the result is a mix. If the left end of the tank has recently yielded better results than right, the tendency for staying there increases. But even if a fish has reason to be satisfied with its end of the tank, it will try the other end from time to time. Gratifyingly, this is what Milinski's stickleback actually do.

Most learning rules operate through reward or punishment. It is obvious that if reward and reproductive success do not match, such learning cannot optimize fitness. Ultimately, the match must rest on innate mechanisms. Sweet tastes better than bitter, for instance, because bitter means poison and sweet means calories, at least to a reasonable first approximation. That's why lumps of sugar were so useful in teaching tricks to *Clever Hans* and countless other horses. But, as noted by Maynard Smith, there is no foolproof system for translating immediate reward into fitness: 'moths fly into candles, human beings become addicted to heroin, and reed warblers raise baby cuckoos.'* While food morsels are *some* measure of success at foraging, they do not take account of the risk from predators, for instance. A pigeon would have little chance of learning what is best in a set-up where the left key, once in a while, opened a sliding door for the laboratory cat.

Exploring new strategies can be dangerous; and repeating every so often a strategy which has not paid off so far is generally costly. And yet such groping is required for any type of learning. One way to reduce its costs is through *gaming*.* This is essentially the simulation of situations involving uncertainties and conflicts, and it is increasingly used as a tool for teaching and training. Economists have devised a number of such war-game-like exercises, which involve two or more decision-makers in complex scenarios. By now more than half of the business schools in the USA have the playing of business games as part of their curricula.*

But the principle of exploring tricky situations by doing *as if* is, of course, much older. Most childhood or adult games contain at least traces of it. To use games for modelling conflicts is itself a time-honoured strategy.

8

Reciprocity and the evolution of cooperation

8

Reciprocity and the evolution of cooperation
Prisoners of a game

And the Devil was pleased, for it gave him a hint
For improving his prisons in Hell.
Coleridge

A 'defect' without fault

A tried and tested way to add spice to a game is to repeat it. Already the first rematch, whether in chess or in tennis, offers added interest. And why stop at one repetition? The periodic return is an essential ingredient of festivals, passion plays, or Olympic games. Play thrives on repetition. Music is, in a way, the art of repetition. (So is, in quite another way, computer programming.) As we have seen in Chapter 4, even a game as exceedingly dull as Heads or Tails leads to intriguing paradoxes, if one only keeps iterating it. Children never tire in their eagerness for repetitions. As for their parents, they watch the reruns of television programs; it is the theme of *Play it again, Sam*, over and over again.

Game theorists, like gamblers and children, can become addicted to iterated games. Their classic example is the *Prisoner's Dilemma*, whose diabolical simplicity has given rise to thousands of scientific publications.* Here is how it goes. Two players are engaged in the game. They have to choose between two options, which we term *Cooperate* or *Defect*. If both cooperate, they can earn three points apiece as *Reward*. If both defect, they get only one point each, which is the *Punishment* for failing to join forces. If one player defects while the other cooperates, then the defector receives five points (this is the *Temptation*) while the trusting cooperator receives no points at all (this is the *Sucker's payoff*). Figure 8.1 encapsulates all this.

How will the rational player act?* By defecting, of course. This is the right choice, no matter what the other player does. Indeed, against a cooperating player, one earns five points by defecting, but only three points for cooperation. Against a defecting player, one earns one point if one also defects, but

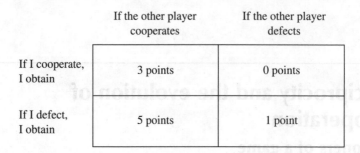

	If the other player cooperates	If the other player defects
If I cooperate, I obtain	3 points	0 points
If I defect, I obtain	5 points	1 point

Fig. 8.1 The payoff for the Prisoner's Dilemma.

nothing at all if one cooperates. Hence, *Defect* is always the best option. The trouble is that the other player, being rational too, thinks along exactly the same lines. As a result, both players end up with only one point each—two less than mutual cooperation would have produced for each of them.

For most people, this conclusion is not very easy to swallow. Indeed, I am happy to report that in experimental tests players frequently prefer cooperation.* In fact, cooperation has such a positive image that defecting makes one feel like a louse. Quite obviously, a good conscience is worth more than the silly difference of a few points.

But that's not playing the game. We must neglect emotions in this thought experiment, and forget about possible ties of friendship and loyalty. We should imagine that the players are two strangers meeting in a dark alley: suspicion reigns, rather than sympathy. And we must keep in mind that the payoff is the only thing that counts. Perhaps it would help to think of some solid gold bars instead of those measly points. (It certainly helps me.) Let us further assume that the transaction is kept strictly anonymous, being arranged through some equally anonymous third party, so that there is nothing personal in it at all.

Do you still feel that the players ought to cooperate? Then try to view the game in an evolutionary setting. Assume that the world consists of two kinds of strategists: defectors and cooperators. Each individual has only one shot at the game, against a randomly drawn opponent. Nobody can know the other's mind. Each point which a player earns corresponds to one offspring, and the offspring inherit the strategy of their parent. Since defectors always do better, their proportion is bound to increase. No matter what we might feel about the right thing to do, the population will inexorably twist away from cooperation.

Game theory helps, as we have seen in Chapter 7, to understand evolutionary biology. But the foregoing argument shows that it can work the other way, too (which is a fitting introduction to a chapter on reciprocity). It shows quite clearly that the Prisoner's Dilemma is not, in fact, a dilemma at all. Defection is the only rational option.

A prison without bounds

If no *Dilemma*, then why *Prisoner's*, by the way? Because the original tale from which the name derives concerned two prisoners accused of a joint crime.* In their situation, to cooperate means to keep silent (that is, *not* to cooperate with the police), while to defect means (if my Chandler serves me well) to sing. If both prisoners remain obdurately silent, they can be kept in confinement pending investigation, but ultimately must be released—say after six months' time. If one of the two prisoners turns state's evidence, he (or she) will get off scot-free as state's witness, while the other prisoner will have to bear the full brunt of the law—five years in the slammer. Thus it seems like a good idea to confess; but if both prisoners confess, both take the rap—two years up the river. They know all this, and yet cannot prevent it.

In the original version, the payoff is measured, not in points, but in length of confinement. It could also be money, or sweets, or prestige. All that matters is the ordering of the payoffs: the Reward for mutual cooperation is preferable to the Punishment for joint defection, but the Temptation for unilateral defection is still better than the Reward, while the Sucker lands, predictably, at the bottom end of the scale. Whenever the outcomes are arranged in this rank order, both players will defect (if they act sensibly, that is) and end up with the Punishment instead of the Reward.

For many reasons, the version with the jail-birds is not applicable to what follows, and therefore we bid them goodbye. (They could be innocent, by the way.) The Dilemma is not confined to prisons. Its horns lurk, for instance, behind *every* transaction. Both parties expect an advantage from the exchange; but this advantage would be greater yet if one's own contribution were omitted, or at least a bit reduced. Usually some kind of authority makes sure that the temptation to defect does not become overwhelming. Education, law, police, and social pressure tend to keep even the most selfish sharpster in line. All this stuff with contracts and 'the party of the first part' sees to it. But if a central authority is missing, the problem becomes acute.

A program for an encore

How can there ever be a cooperation between competitors in the absence of conscience or constraint? Well, possibly through the prospect of repetition. It is obvious that traders in the habit of defaulting on their partners will soon find no more partners to default on. Some hope of further trading is certainly a strong incentive for fair deals.

So let us check whether a *repeated* Prisoner's Dilemma game can do the trick and lead to cooperation. If the number of repetitions is known in advance to both players, we will be quickly disappointed. Indeed, the last game in the series is a simple one-shot Prisoner's Dilemma again, and we know its outcome already: no cooperation. This will be unaffected by good or bad behaviour in the previous rounds. What's past is past; neither gratitude nor a burning desire for revenge should divert us from the aim of maximizing our score. Hence the outcome of that last game is settled, and we may as well forget about it. What about the game before? Obviously, the same argument holds. And so on. We can shorten the problem step by step, like the Marx brothers trimming a moustache by little 'snoops', until nothing remains in the end.* No cooperation, all the way down.

It is the *expectation* of further dealings which makes cooperation so alluring. The end of the interaction must not be known beforehand. There should always be *some* probability of a further round. Let us denote this probability by w and assume for the sake of simplicity that it remains constant. (The case where the readiness for a further encounter depends on the outcome so far is obviously of interest too, but we defer it for the moment.) If w, for instance, is 90 per cent, there will be 10 rounds of the game *on average*. There could actually be 20 rounds, or only one. But no matter how many rounds have been played so far, the probability of a further round remains unflaggingly 90 per cent.

A *strategy* for such an Iterated Prisoner's Dilemma is a program prescribing in each round whether to defect or to cooperate. This decision may depend on the number of steps (for example *defect in every third round; otherwise, cooperate*), and it may depend on chance (for example *throw dice, and defect if a Six shows up*). The decision can also depend on the previous outcomes (for example *defect if the other player has so far defected twice as often as you*). But it cannot depend on the *future* outcomes, of course. No crystal balls are allowed in this game. No player knows the other player's program.

And here's the rub. In the simple Prisoner's Dilemma, a knowledge of what the other player is going to do is absolutely irrelevant. No matter what, the best reply will always be to defect. In the Iterated Prisoner's Dilemma, this is different, at least if w is sufficiently large. There is no strategy, now, which is always the best reply.* If, for example, the strategy of the other player is to cooperate right up to my own first defection, and from then on never to cooperate again, I would be foolish ever to defect. Indeed, a defection will yield an immediate gain in the present round, but smash any chance of future mutual cooperation. If w is large, then so is the expected number of forthcoming rounds. If they are filled with defections, they carry a loss which cannot be compensated by the short-term gain in one single round. (It does not matter, incidentally, that the other player loses still more. All I have to worry about is my *own* total score.) If, on the other hand, the strategy of the other player consists of always defecting, then my best reply is also to always defect. In these two cases, two different strategies maximize my payoff. There exists no universally valid best strategy, therefore, if the probability of a repetition looms large. (If it is small, on the other hand, then the future interaction is negligible and the Iterated Prisoner's Dilemma is not iterated enough to differ from the simple Prisoner's Dilemma: defection will be best, come what may.)

What is one to do, then, if the probability of an iteration is large and the future casts a long shadow? One cannot pry into another's brain. The only reliable guideline is that opponents, too, are out to maximize their payoff. Is this enough to yield a solution? The question proved to be far from easy, and game theorists kept arguing about it for decades, until 1979, when a young American professor of political science named Robert Axelrod had the glorious idea of inviting them down to a tournament.* It was conducted by mail.

A titter at the tournament

Fifteen participants sent in their programs, and Axelrod turned them loose in his computer. As in a soccer league, each program was matched with each other one; on top of that, it had to measure up against a copy of itself, and against the random strategy which opts at every step with equal probability to defect or not. And when the dust had finally settled after the fray . . .

Well, in the old romances of chivalry it is usually an apparently dumb and hitherto overlooked simpleton who emerges as the unexpected winner. That's what happened here too. The shortest and simplest of all programs won hands down. It was called *Tit For Tat* and consisted of cooperating in the first round and then doing whatever the other player had done in the previous round. First cooperate and then imitate, in a nutshell.

In those romances of chivalry it invariably turns out that the modesty of the young knight hides a lineage of the highest nobility. *Tit For Tat* has indeed a superb pedigree. It was submitted by the Canadian Anatol Rapoport, a former concert pianist and one of the grand old men of game theory. Rapoport had pondered on the Prisoner's Dilemma longer and deeper than anyone else.* He had a special incentive for this. Over several decades, he had been strongly committed to peace movements. The arms negotiations between the two superpowers were the most spectacular example of a Prisoner's Dilemma. Obviously both parties would have profited from bilateral disarmament; but each would have gained still more from the unilateral disarmament of its rival. The Prisoner's Dilemma was not merely a game for Rapoport. And when he entered *Tit For Tat*, he knew what he was betting on.

Its being backed by Rapoport is just about the only thing, however, which is *not* surprising about the tournament winner. This can best be seen from the fact that *Tit For Tat* won not a single match. In fact it never wins. You can earn with it as many points as your opponent, or less, but certainly not more. Indeed, a *Tit For Tat*-player is *nice* in the sense of never being the first to defect. If a run of mutual cooperation is ever broken, it is because the other player defects—thereby moving ahead in the count. In the following round the *Tit For Tat*-player retaliates. If the other player defects again, the lead remains the same; otherwise, the difference will be cancelled. But as soon as this happens, the *Tit For Tat*-player is in cooperation mode again. A player with this strategy will never draw ahead. And after every round the game can come to an end. If this happens after a defection by the other player, the *Tit For Tat*-player no longer has an opportunity to catch up. During the entire race, *Tit For Tat* is never out in front (but never very far behind either).

A chess player who never wins a single game cannot conceivably end a round-robin tournament ahead of all the rivals. Chess is a zero-sum game (or more precisely, a constant-sum game): the gain of one player is the loss of the other. The Prisoner's Dilemma, however, is not a zero-sum game. Nor, incidentally, is it a partnership game with both players sharing the payoff. It is something in between. The interests of the players are neither diametrically opposed nor identical. In a round-robin tournament, each player meets each other once: if there are 10 players, say, there will be 45 games altogether. A win, at chess, yields one point, a defeat zero, a draw half a point. Thus there will be 45 points dished out during the tournament: whoever wants some will have to wrestle them away from the adversary. On average, each player gets 4.5 points. One cannot do better without winning at least once. In the Iterated Prisoner's Dilemma, however, the total amount of points depends on the strategies of the players. My doing well does not necessarily preclude the other player from doing well too.

Cogito zero sum

Most parlour games are zero-sum games. For this reason, as we have seen, the zero-sum situation held centre stage during the early years of game theory. But most of real life is not zero-sum. This is a truism, but it seems to run against deeply ingrained prejudice. It is often argued, for instance, as if wealth were something constant; that it has been taken away from the poor classes by the millionaires, and that it should be redistributed. But attempts to do so have failed. Most social and biological interactions are closer to the Prisoner's Dilemma than to poker. Even if the interests conflict, they usually do not clash head-on.

It is true that the very best outcome of an Iterated Prisoner's Dilemma would be to defect consistently against an inveterate sucker. But this is certainly too much to hope for. It is more realistic to aim for an unbroken run of mutual cooperation. This yields a reward in each round, and guarantees a score which should be viewed as a benchmark for the total payoff. The other party, in such an interaction, does equally well. Actually, there is little chance of doing consistently better than the benchmark.

(A technical aside: this would be different if the Temptation amounted to 10 points instead of five. In this case, my partner and I could decide to defect in turns, and share our net income to mutual advantage. We would get 5 points each, which is better than a Reward of 3 points. But since it is generally felt that this would be carrying cooperation too far, game theorists add a rule to the Prisoner's Dilemma which kills this option effectively: Reward is to be larger than the mean of Temptation and the Sucker's payoff.)

In Axelrod's tournament, the benchmark was never exceeded *on average*. Eight of the 16 entries were nice in the sense of never defecting first, and these eight did consistently better than the rest. There was even a substantial gap between the earnings of the two groups. Whenever two nice strategies came to meet, they enjoyed an unbroken run of mutual cooperation. The less nice strategies, by contrast, got embroiled in costly vendettas again and again. Every attempt to outwit and exploit the other player by throwing in an occasional defection ended in dismal failure.

While the nice strategies all did equally well against each other, they differed markedly in their scores against exploiters. The advantage of *Tit For Tat* lies in it being quick to retaliate and quick to forgive. The strategy which never forgives, but defects in a relentless run from the first defection of the opponent onward, was the entry which finished worst among all nice strategies. Everlasting wrath can obviously be a very costly emotion. And, to remain sententious (a temptation which is not to be resisted in this chapter): swallowing one's anger is not a good thing either. Some wily programs in this tournament probed the forbearance of their fellow-players by occasional defections. But a *Tit For Tat*-strategist retaliates immediately after every provocation and can quickly convert most probers to cooperation. The swift and clear answer signals that everything will be returned forthwith. It leaves the decision about what it is going to be entirely to the other player. *Tit For Tat* manipulates by means of its own transparency. Which shows that extroverts are favoured in life.

A good many of the tournament entries consisted of variants of *Tit For Tat* whose trimmings and elaborations aimed at the occasional dash for a few extra points. Such cunning refinements were to no avail. What Rapoport had understood best of all was that in the Iterated Prisoner's Dilemma, envy does not pay. One's own well-being is what matters. An egoist should think of nothing else.

Be robust or bust

But of course even *Tit For Tat* is not the best of all possible strategies. In his review after the tournament, Axelrod easily found strategies which would have done better than *Tit For Tat*, had they been entered. *Tit For Two Tats* was one of them—a strategy which tolerates isolated slips, and only defects if the other player has defected in the previous *two* rounds.

Axelrod published his analysis and called for another tournament.* A *sequel*: isn't this what every Hollywood producer—or bestselling author—likes to add to a success? By now, interest in the Prisoner's Dilemma was no longer confined to the circles of game theorists. Economists, mathematicians, engineers, biologists, and a good many computer freaks threw themselves into the *melée*. Sixty-two programs were submitted.

Because the news of the first tournament had made the rounds, it was to be expected that the new contest would be considerably more demanding. *Tit For Tat*, again submitted by Rapoport, was going to find itself in deep waters indeed. John Maynard Smith, for instance, not one to shrink from the simple solution, entered *Tit For Two Tats*. But while this strategy would have won the first tournament, it finished only twenty-first in the second. One would expect *Tit For Tat* somewhere even lower down on the list. But no. Amazingly, *Tit For Tat* also won the second tournament. And quite properly so: a sequel should offer new adventures, but not, one hopes, a new hero.

Generals are reputed to always prepare for yesterday's war. This is called 'drawing lessons from history'. In much the same spirit, the participants at the second tournament drew their lessons from the first. But the many lessons which had been learned cancelled each other out, somehow. Rapoport had not drawn any lesson. Incorrigible, if you like. There was obviously not much left for him to learn.

If all contestants of a tournament were known, one can doubtlessly find some strategy for improving upon *Tit For Tat*. But the contestants are not known. *Tit For Tat* is not a tailor-made strategy, but rather one off the shelf. Still, it fits pretty well in most cases. *Tit For Tat* is robust.

The fact that the contestants are not known need not keep us from speculating, by the way. Among the infinitely many strategies, many, if not most, are just plain stupid, so some should be more likely than others. Rapoport, for example, knew how to make an educated guess.

To guess means to select. So maybe one should think in terms of natural selection. It helped in convincing us that in a one-shot Prisoner's Dilemma, cooperation does not stand a chance. In the repeated Prisoner's Dilemma, selection could weed out a lot of strategies. This narrows down the field of possible encounters, at least after an initial phase of trial and error.

Hence Axelrod devised a kind of evolving competition. The participants now form a population whose composition changes with each repetition of the tournament. The successful strategies will multiply, while their hapless rivals will fade out. But if the field of participants changes, so will the recipe for success. What is good today need not be good tomorrow. Once more, we meet a feedback loop: success determines the composition of the field, and the composition determines success. It is not easy to predict where this may lead. Axelrod simulated the effect of natural selection with his computer. There was a tournament in every generation—an iteration of the Iterated Prisoner's Dilemma—and the frequency of a strategy increased in proportion to its payoff in the previous generation.

As expected, *Tit For Tat* took the lead right away—this is, after all, the prize for winning the tournament in the first generation. Somewhat less obviously, *Tit For Tat* went on extending its lead. Even after 1000 generations (that's when Axelrod finally blew the whistle), its rate of increase was still the highest. There were other strategies which initially showed promise, but then lost their breath, trailed behind, and ultimately left the track. In particular, this was the fate of all strategies whose success depended on ruthless exploitation of others. For these victims were soon eliminated, and then there was nothing left to exploit. On the other hand, whenever a *Tit For Tat*-player did well in an encounter, the other player did at least as well, and came out strengthened by the trial. This meant more favourable encounters of this sort in the next round. Such a success is of an enduring kind: it grows upon itself. Some strategies are their own worst enemy, but *Tit For Tat*-players have nothing to fear from that angle. On the contrary, they can never do better than against each other. Indeed, they will enjoy an unbroken run of mutual cooperation against any nice player they meet.

A one-way street to happy returns

Dim are the prospects, however, of a *Tit For Tat*-player in a population of inveterate defectors. In such a snake-pit, the defector will always do slightly better. (The difference is due to round 1, where *Tit For Tat* is caught unawares.) *Tit For Tat* can only get the upper hand if its frequency is above some critical level: for then it will experience some interactions with other *Tit For Tat*-ers, and a few of these episodes of cooperation will be enough to compensate for the slight disadvantage against all-out defectors. Indeed, these defectors earn considerably less than the benchmark score, whether they meet their own kind or *Tit For Tat*. Hence there is an effective threshold for mixtures of *Tit For Tat*-players and all-out defectors. For instance, if $w = 90\%$, *Tit For Tat* draws ahead as soon as more than 6 per cent of the population use it. Only the initial phase of entering into a world of exploiters is fraught with danger: to overcome it, *Tit For Tat*-players have to appear in a cluster—as a family, for example. Once the threshold is crossed, there is no looking back.* The number of *Tit For Tat*-players grows, and every increase boosts the success of their strategy.

This pattern of family clusters invading makes sense. Kinship facilitates cooperation. The offspring of your relative is likely to carry many of your genes, and can be viewed as a watered-down version of your own offspring. Remember that in our games, the payoff is number of offspring. In a game with a relative, therefore, every point scored by your coplayer yields, depending on the relatedness, a fraction of a point for yourself. The payoff is rigged: you have a vested interest in your partner. The Mafia was not the first to discover that sheer selfishness may dictate some amount of cooperation within the family.

It is of no small importance, of course, to be able to recognize relatives. Now if one cooperates within a family, the reciprocation of cooperation can be used as a cue for relatedness. It is therefore but a small step from the program *Cooperate with relatives* to the program *Reciprocate*. The way is open for *Tit For Tat* to spread, and to outgrow its family circle.

This issue of relatedness and reciprocation has come up in a study of vervet monkeys responding to calls for support.* The monkeys seem to recognize each other's voices. Their calls for assistance were tape-recorded, and later played on a loudspeaker hidden in a bush. It isn't that a vervet monkey, upon hearing such a call, immediately rushes to the spot. If it so much as looks towards the bush, this is considered a positive response already. Now it turned out that if a monkey had been groomed by the caller some short while ago, it was much more likely to respond. The grooming received from the comrade apparently boosts the willingness to assist in case of need. This reciprocation only took place, however, if the monkeys were *not* related. Within the family, responsiveness was higher than average, but remained unaffected by recent grooming. It almost looks as if the two rules *Help your helper* and *Help your relative* were used only separately. Acts of kindness appear to elicit less gratitude among kin. They seem to be taken for granted. This literally sounds familiar, doesn't it?

The good news in all scenarios for the emergence of reciprocation is that the scales are weighted in the game. A one-way street leads from defection to cooperation. This seems surprising at first glance. A single *Tit For Tat*-player among all-out defectors and a single defector among *Tit For Tat*-players are both in pretty poor shape. But while *Tit For Tat*-players can invade if they alight in a clique, all-out defectors cannot. On the contrary, every other defector they meet is a nail in their coffin, robbing them of that slight edge in the first round. In this sense the threshold works as a kind of ratchet, favouring the evolution of cooperation. (We see, incidentally, that by matching each strategy against a *doppelgänger* of itself, Axelrod influenced the outcome of his tournament in favour of cooperation: two nice players are natural allies, two defectors certainly not.)

The meek shall endanger the earth

It is not exploiters who can seriously threaten a population of *Tit For Tat*-players, but those who let themselves be exploited. A player consistently opting for cooperation, no matter what happens, does neither better nor worse than the rest in a *Tit For Tat*-population: mutual cooperation wherever one looks. Natural selection will neither frown nor smile upon such players. In a neutral equilibrium of this type, random drift takes command. Frequencies dawdle aimlessly up and down. Sooner or later, a considerable fraction of the population will consist of individuals unable to react, and committed to unflagging cooperation. *Tit For Tat*, after all, demands a modicum of memory, as well as the ability to take notice of the other player's behaviour and to react with promptitude and dispatch. In an environment of certified nice guys, there seems no harm in dropping these sterling qualities and switching to the less expensive mode of all-out cooperation. No harm, that is, in the short run. But individuals unable to retaliate are just what exploiters are waiting for. They lurk in the bushes. Even a small number of born victims is enough to allow all-out defectors to invade, at least up to the point when *Tit For Tat* rallies round. But if *Tit For Tat* has become too rare, there will be no rallying round, and all-out defectors take over.

It is dangerous to lose your resistance against exploiters. If you allow yourself to be exploited, you will not just burden your own account; by supporting exploiters, you threaten the whole community.

It's pretty hard not to start moralizing at this point. So let's have a stab at it. It would of course be rather silly to attempt to reduce all human interactions to the Iterated Prisoner's Dilemma, or to negate the role of superior authority in civilized communities. But with all due restraint, it is worth pointing out that the brutally simple principle of paying back in kind leads to cooperation in a society of egoists, while the apparently higher summons to dispense with reprisals undermines such cooperation.

The meek turning of the other cheek forbids requital upon provocation, and destroys the basis of reciprocity. There is a streak of anarchy in Christian doctrine. In fact, the overstrung teachings of the gospels were probably not meant to serve as the foundation of an enduring order. For early Christians, the end of the world was imminent, and the prospect of reward in the next life modified, of course, the payoffs in the Prisoner's Dilemma.

The harsh law of retaliation seems to have been the foundation stone of many, possibly all, stable societies. 'An eye for an eye, a tooth for a tooth': such grim transactions ensure the emergence of fair trade. But as we shall see next, the maxim of retaliation is not wisdom's last word.

White noise and black marks

Our little moralistic digression started off with the remark that a *Tit For Tat*-population can be subverted by a fifth column of all-out cooperators ushered in through random fluctuations. This did not happen in Axelrod's computer experiments (at least, not in those described so far) because they left nothing to chance. But chance, of course, has to play a part in every decent model of a biological community. It could make itself felt, for instance, through mutations or random sampling. But it could also be part of the strategy or the interaction itself. While one can reasonably expect that a computer program will be carried out without fault, this is asking too much from the behaviour of an animal. Any life-like interaction is bound to be fraught with misunderstandings and errors. There will be unpredictable fluctuations from the environment. There will be mistakes in perceiving what the other player does, and mistakes in carrying out one's reply. The trembling hand, the fuzzy mind, the shaky ground: it adds up to crossed wires. After all, communication is essential in this game. Opting for cooperation or defection acts as a *message* in each round; and, like every exchange of information, such a message is subject to perturbations.

Even the smallest background noise can come very costly to a *Tit For Tat*-population. In the long run, a mistake is bound to occur, which unilaterally stops cooperation. The other player replies to this defection, thereby provoking a defection in turn. Thus the echo of an erroneous signal will carry back and forth. The players take turns in defecting and cooperating. This excruciatingly silly see-saw of recriminations will only be stopped by a further mistake, which can restore cooperation, but which may equally well lead to a run of *mutual* defections. In any case, mistakes cause feuds, and feuds do not pay.

To verify this, Axelrod's tournaments were repeated to the accompaniment of random perturbations.* An error rate of 1 per cent still left *Tit For Tat* in its leading position, but with 10 per cent it fell back. Actually *Tit For Tat*'s high sensitivity to noise is less of a handicap in the mixed population of the tournaments than in a population of copies of its own. The closer *Tit For Tat* comes to dominating the field, the more expensive errors prove to be. The same inflexibility which frequently converts others in no time at all to grovelling cooperators leads against its own like-minded clones to endless runs of reverberating defections. A little sprinkling of mistakes, and the average payoff in a *Tit For Tat*-population drops by 25 per cent, from 3 to 2.25 points per round. This is a very poor showing. The brainless random strategy which defects or cooperates in each round with equal probability does no worse.

The obvious ploy for breaking up the vicious circle of grim retaliation consists of being ready, on occasion, to let bygones be bygones. We can even compute the optimal measure of forgiveness, at least when the game is very likely to be repeated.* Let p and q denote the probabilities of cooperating in the next round, given that the other player has just opted for cooperation or for defection respectively. Each pair (p,q) of numbers between 0 and 1 then corresponds to a randomized strategy, a strategy which replies to the other player's move, not with a clear-cut determination one way or the other, but rather with a stronger or weaker tendency to cooperate. Each such strategy (p,q) is given by a point in the unit square (Fig. 8.2). *Tit For Tat* corresponds to the limiting case of perfect reciprocity: $p = 1$ and $q = 0$, the south-east corner of the square. The purely random strategy $p = q = 50$ per cent lies smack in its centre. Always defecting means $p = q = 0$, and corresponds to the south-west corner, while the opposite north-east corner denotes the strategy which always cooperates: $p = q = 100$ per cent.

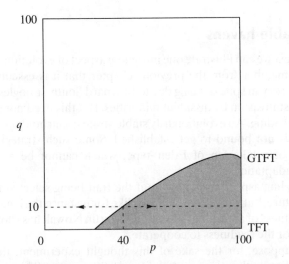

Fig. 8.2 The (p,q)-strategy space. p is the probability of cooperating if the other player cooperated in the previous round. q is the probability of cooperating if the other player defected in the previous round. The lower left-hand corner $(0,0)$ is that of unconditional defectors, the lower right-hand corner $(1,0)$ that of *Tit For Tat* (TFT). The upper right-hand corner $(1,1)$ is that of unconditional cooperators. In the cooperation-rewarding zone (shaded area), it pays to increase p and q, that is, the readiness to cooperate. *Generous Tit For Tat* (GTFT) is the optimal strategy for the population, for very small error probabilities.

If $q = 10$ per cent is fixed throughout the population, the value $p = 40$ per cent is evolutionarily stable: if all members of the population use it, no dissident minority can invade. But if all members use a p-value which is larger than that, then minorities with still larger p can take over, whereas minorities with smaller p (for instance, 40 per cent) cannot. Thus the trend is *away* from evolutionary stability.

There exists a neighbourhood of *Tit For Tat* where every increase in the readiness to cooperate will be favoured by selection, which means that mutants with higher p- or q-values will spread whenever the population is in this cooperation-rewarding zone.* Beyond this zone, deviations to *lower* values of p or q will flourish, and the average payoff decreases. Therefore, only the strategies in the cooperation-rewarding zone are immune to exploiters. Among these immune strategies, the *Generous Tit For Tat* strategy (which is given by $p = 1$ and $q = 33$ per cent if the noise-level is small) is optimal: its payoff is maximized. Higher values of q would provide the population with even better scores, of course, but would allow defectors to cash in.

So the way to deal with the risk of occasional misunderstandings is never, as far as possible, to forget a good turn, but sometimes (not always!) to forgive a bad turn. It is in one's own interest to develop a high sense of gratitude and a limited dose of tolerance.

Unreachable havens

In this context we can illustrate one intriguing aspect of evolutionary stability. We may remember from the previous chapter that it is essentially a conservative notion, in spite of being due to Maynard Smith. It singles out what is *proof* against invasion by dissident minorities. But this need not mean that it is good at invading. An evolutionarily stable strategy can hold on, once established, but is not bound to get established. Some such strategies are configurations of a 'Garden of Eden'-type, which cannot be arrived at by successive adaptations.

This puzzling aspect may show up if the trait being selected is not of an *either/or* nature, but varies continuously, like for instance the length of a horn or the brightness of a warning coloration.* Martin Nowak has shown that this also holds for the readiness to cooperate.*

Let us suppose, for the sake of this thought experiment, that all individuals retaliate with a given probability, say 90 per cent. This means that q, the likelihood of cooperating after a defection of the adversary, is 10 per cent throughout the population. In this case, the value $p = 40\%$ happens to be evolutionarily stable. This means that when all members of the population cooperate with a likelihood of 40 per cent after a cooperative move of their coplayer, then a minority with a different p-value will do worse than average, and hence vanish. But suppose now that the common value of p, in the population, were different from the evolutionarily stable 40 per cent, however slightly. Suppose, for instance, that it is 41 per cent. In this case, the population is in the cooperation-rewarding zone. This means that a minority with a p-value of 43, or 60, or 80 per cent, can successfully move in. But a minority with a p-value lower than 41 per cent will be selected *against*. In particular, the evolutionarily stable value 40 per cent cannot invade.

Conversely, a population where all members had a *p*-value of 39 per cent, say, would be in the zone where less cooperation is favoured. A minority with *p* = 37 per cent could spread; but a minority with a higher *p*-value than 39 per cent could not. Hence, the trend is *away* from the evolutionarily stable state! The 40 per cent act like a watershed. If *p* is smaller, it evolves towards smaller and smaller values; if *p* is larger, it will grow larger and larger. In the long run, the value of *p* will either be 0 or 100 per cent.

Some evolutionarily stable strategies, then, are unattainable. Even if, by an inconceivable fluke of fate, the population started out with all individuals using precisely the ideal value *p* = 40 per cent, a slight variation of the environment—causing a barely noticeable shift in the payoff for mutual cooperation, for instance—would somewhat alter this ideal value, so that the population would find itself on one side or the other of the watershed and start on its fall from grace, irreversibly moving away from 40 per cent. Even if the environment returned to normal, this would not stop the slide.

Strangely enough, the situation with *q* is just the opposite. If all members of the population used *p* = 90 per cent, say, then the value *q* = 35 per cent is evolutionarily stable, and also attainable, in a sense unavoidable even. If the *q*-value were higher, selection would tend to make it smaller; if it were smaller, it would make it higher. This stabilizing sort of regulation would safely steer the population into a haven which cannot be invaded.

Crowd control

Even an evolutionarily stable strategy can be overcome if *several* strategies invade simultaneously. This is illustrated by a scenario due to R. Boyd and J. P. Lorberbaum.* A *Tit For Tat*-population cannot be invaded by *Suspicious Tit For Tat*, the strategy which defects in the first round and then reciprocates. Indeed, the initial move of such a strategy causes an end-less series of alternating defections. *Tit For Two Tats*, on the other hand, tolerates an isolated defection. Since *Tit For Two Tats* and *Tit For Tat* do as well, against each other, as against themselves, any mixture of these two is neutrally stable. But because an invading minority of *Suspicious Tit For Tat* cooperates with one but not with the other, the proportion of *Tit For Tat*-players will decrease. Hence a community of *Tit For Tat*-ers can be invaded by a coalition of the two other strategies entering hand in hand, so to speak.

Tit For Two Tats is a generous strategy, and therefore well protected against background noise. Indeed, it forgives isolated faults; and the prob-ability of two mistakes in a row is very small. But the trouble with *Tit For Two Tats* is that one can easily figure out how to exploit it at no risk, while a strategy which retaliates after 50 per cent of all defections, for instance, leaves a prober much more in the dark.

In fact, as soon as *several* strategies are around in sizeable numbers, the outcome may be difficult to predict. Selection can lead to stable mixtures, or to cyclic revolutions (A beats B, B beats C, C beats A), to periodically or chaotically oscillating compositions.* All this happens even within the range of those strategies which remember only the last round. *Tit For Two Tats*, for example, does not belong to this tribe.

In a mixed crowd of players, it is extremely hard to decide what to do. One would have to know a *representative sample* of the population in order to get bearings.* It is conceivable that the participants at Axelrod's tournaments constitute a representative sample of game theorists. But to extrapolate from there to lower forms of life would be rash. One should proceed more methodically.

Unfortunately, the array of strategies for the Iterated Prisoner's Dilemma is so huge that it cannot be sampled by statistical means. It has first to be reduced drastically. One way is to look at (p,q)-strategies only. In spite of their simple build, these strategies cover a broad spectrum of cooperative, defective, fully random and reciprocal behaviour, and their probabilistic nature reflects the fuzziness of simple social interactions, where memories are short and decisions uncertain.

We may start out with 100 such strategies, for instance, randomly scattered over the unit square, and watch them thrash it out.* Since we do not want to bias the result, all these strategies should initially be equally frequent. But some will do better than others, and spread from generation to generation, while others will dwindle.

In almost all such simulations, the evolution proceeds towards *Always defect*, the strategy where both p and q are 0 (Fig. 8.3). This follows because many strategies from the initial random sample have relatively high q-values, and hence are likely to cooperate even if their coplayer has defected. These strategies cannot firmly retaliate against exploiters. With such a rich menu of 'Suckers', it pays to defect. In the end, all that remains from the original population are those strategies which are closest to the defector's corner $p = 0$ and $q = 0$.

The outcome changes dramatically if one of the initial strategies is *Tit For Tat*, or very close to it: its probability p of cooperating after a cooperative move of the coplayer must be almost 100 per cent, and its probability q of cooperating after a defection almost 0. The first phase—which lasts for some 50 generations or so—is practically indistinguishable from the previous run. The strategies in the defector's corner grow rapidly. The 'Suckers' are wiped off. *Tit For Tat* and all other reciprocating strategies (with p almost 100 per cent and q almost 0) seem to have disappeared. But an embattled minority hangs on by the skin of its teeth. The tide turns when the 'Suckers' are so

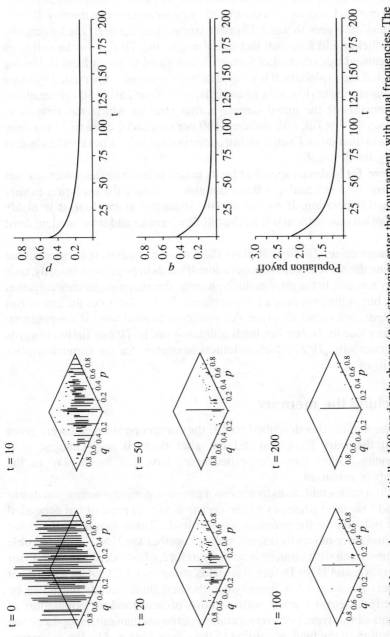

Fig. 8.3 Nasty take-over. Initially, 100 randomly chosen (p, q)-strategies enter the tournament, with equal frequencies. The strategies closest to all-out defection $(p = q = 0)$ take over. The average values of p and q as well as the overall payoff in the population decrease quickly.

decimated that exploiters can no longer feed on them. That's when the retaliators begin to rally round and inch their way up. It is impressive to watch them come back, slowly at first, but gathering momentum on the way. It is the exploiters' turn now to wane. Disaster strikes their corner. One by one, the nasties flicker and fade out. But interestingly, the *Tit For Tat*-like strategy which caused this reversal of fortune is not going to profit from it. Having eliminated the exploiters, it is robbed of its mission and superseded by more generous strategies (Fig. 8.4). Eventually, after some 250 or 300 generations, what remains of the initial sample is that strategy which was nearest to *Generous Tit For Tat*, with p almost 100 per cent and q close to 33 per cent. Evolution then stops. Even if we introduce occasionally 1 per cent of another strategy, it will vanish.

Tit For Tat is almost specified by its police role: strategies which are not *very* close to $p = 1$ and $q = 0$ are not able to induce the evolution to veer towards cooperation. If we select 100 strategies at random, it is highly unlikely that one of them will fit the bill. We have to add it by hand, in most cases.

To summarize: *Tit For Tat* plays the role of a catalyst. It is essential for triggering the move towards cooperation. It needs to be present, initially, only in a tiny amount; in the intermediate phase of the 'reaction', its concentration is high; but in the end, only a trace remains. *Tit For Tat* is not the aim of this movement, but rather its pivot. An evolution twisted toward cooperation, and hence due to *Tit For Tat*, leads ultimately not to *Tit For Tat* but towards more generosity. *Tit For Tat*'s strictness is salutary for the community, but harms its own.

Stretching the memory

In all the simulations described so far, the competing strategies were given right at the start. Evolution did not alter them: it only changed their frequencies. Other computer experiments have left more play to the creativity of evolution.

The first such simulations by means of *genetic algorithms* were again due to Axelrod.* He used strategies whose decisions to cooperate or not depended on the outcome of the previous *three* rounds. Three moves by the other player and three moves by oneself make altogether for $2^3 \times 2^3 = 64$ possible outcomes. Each such strategy is a list, therefore, of 64 decisions. With C for Cooperation and D for Defect, this yields strings like CCCDCD ... CDC of 64 letters (actually a few more, since the first three rounds must also be specified). Axelrod started with a more or less random population of programs of this type. For every iteration of the tournament, he updated the composition of the field, according to the scores obtained by the programs. He also threw in occasional mutations (the random change of a letter in the

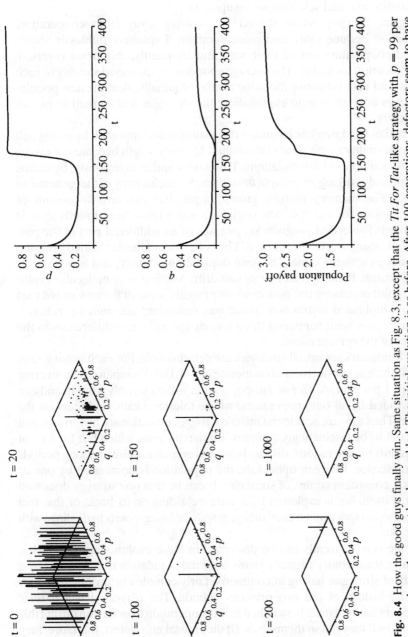

Fig. 8.4 How the good guys finally win. Same situation as Fig. 8.3, except that the *Tit For Tat*-like strategy with $p = 99$ per cent and $q = 1$ per cent has been added. The initial evolution is as before. After 100 generations, defectors seem to have taken over. But after 150 generations, the strategy close to *Tit For Tat* recovers. After 200 generations, the defectors are gone. But now, *Generous Tit For Tat* moves in and supersedes *Tit For Tat*. The overall payoff first drops, then rises sharply.

string) and recombination (the crossover of two strings). Thus he generated variability, and had selection acting upon it.

Again, the population started by evolving away from cooperation. Defection became more and more common. Exploiters ruthlessly shouldered cooperators out of their way. But frequently, this trend reversed: reciprocation took over. The average cooperation increased steadily in such cases, and rose far above the value displayed initially. And the new population was now immune to exploitation: the strategies had learned to punish defectors.

Kristian Lindgren has extended this evolutionary approach by giving still more flexibility to the genetic algorithm. Memory length becomes an evolutionary variable in his simulations. This works, rather ingeniously, by means of randomly striking *gene duplications* which attach a copy of the 'genome' to itself.* The memory thereby grows longer, but the surplus amount of information—for example, the coplayer's move from three rounds ago—is not used. The coded program keeps track of an additional part of the past without basing any decisions on it. The behaviour of the duplicated program remains exactly the same. The gene duplication is 'silent', and hence selectively neutral. It can spread by random drift. And just as in molecular evolution, point mutations can now randomly modify some of the new or old part of the duplicated instruction. What was redundant becomes operational: differences in what happened three rounds ago will cause differences in the choice of the current move.

In Lindgren's set-up, all strategies are deterministic. For each history, they specify a clear-cut decision to cooperate or not. But this is blurred by an error rate of 1 per cent. A *Tit For Tat*-population suffers terribly from it, and can be invaded by all-out cooperators, whose tolerance calls exploiters on the scene. Thus the rare accidental mistakes trigger evolutionary histories which turn out to be fascinatingly complex. Lindgren's runs, which last for tens of thousands of generations, display bouts of feverish turmoil and long periods of quiescence. Between upheavals, the population is dominated by one or several coexistent strains of strategies. It can be that one strategy does well against itself, but is exploited by a 'parasite' riding on its back; or that two strategies do well against each other, in spite of losing points in dealings with their like.

Some general trends can be discerned in these evolutionary chronicles. Usually, the memory capacity grows until the population is dominated by a family of strategies having in common a fairly complex behaviour based on the full history of the two previous rounds. The players defect if their opponent has unilaterally switched from cooperation to defection during this period, or if they did so themselves. (If they mistakenly defect, therefore, they defect right again, as if anticipating an ill-tempered reply from their partner.) On the other hand, the players cooperate if they both cooperated in the

previous two rounds, or simultaneously switched from defection to cooperation. And they also cooperate if they both *defected* in the previous two
rounds. This last rule seems rather odd, but it guarantees that two players
equipped with such strategies will do well against each other in the long run. If
one player accidentally defects, both players defect twice in a row and then
resume cooperation as if nothing had happened.

At first glance, this seems a most unlikely behaviour. But on second
thought, aren't domestic quarrels often enough resolved in exactly this way?
A brief thunderstorm, and the atmosphere is cleared.

All these computer simulations carry the same, highly uplifting message.
Cooperation evolves even in a totally selfish population. Envy and greed
appear futile. It does not pay to do another down. Reciprocity flourishes in a
variety of environments, and it even acts to create an environment to its taste.
It is a self-promoting policy.

A stickler for fairness

With so much support from cyberworld evolution, we may reasonably conjecture that the Iterated Prisoner's Dilemma has something to tell us about
cooperation and reciprocity in wetware biology. Ever since a path-breaking
paper by Axelrod and Hamilton,* this aspect of sociobiology has indeed
attracted much attention.

A particularly elegant experiment was devised by Manfred Milinski,
whose stickleback enjoy a well-earned reputation for keeping abreast of the
newest trends in animal behaviour.* If stickleback notice a pike in their
vicinity, they approach it gingerly (only up to a respectful distance, of course)
and take a good look at it. This is called *predator inspection*, and supposedly
provides the fish with some clues about the current mood and motivation of
the pike. In the previous chapter, we saw guppies inspecting their predators
too. It is obviously useful to know the enemy. Field Marshal Montgomery, for
instance, kept a picture of Rommel on the wall of his caravan. The stickleback
take a somewhat greater risk. This risk, however, is considerably reduced if
the stickleback approach in pairs, as they frequently seem to do. The danger
is reduced by more than half: the efficiency of a predator drops drastically
whenever two targets compete for its attention. As long as the two stickleback approach together or take the lead in turns, one can speak of cooperation. But if one of them consistently hangs back and gains its information
about the pike by waiting in the wings while the fellow stickleback sticks its
neck out, this is a clear case of defection. Milinski had a single stickleback
confronted with a pike (a dummy pike, as a matter of fact), and used a mirror
to make the stickleback think that it had a companion. Depending on the
position of the mirror, the fake companion kept either abreast or a couple of
inches behind. In the former case, the deluded stickleback usually dared to

move a bit closer to the pike—a strong hint for a strategy based on reciprocity.*

These stickleback have discovered a trick to turn a seemingly hopeless one-off Prisoner's Dilemma into an iterated (and hence promising) one. It consists of dividing the interaction into lots of small steps. The partial treaties of disarmament talks operate on a similar principle.

Sometimes, the reciprocating moves are divided into unequal parts. In groups with a strong hierarchical order, for instance among chimpanzees, it has been observed that the subordinate has to accumulate many small acts of assistance, like grooming, while the dominant individual repays only on rare occasions, but usually with an important service, like helping in a fight.*

Asymmetries between the two players of a Prisoner's Dilemma game can be more pronounced still. This holds, for instance, for helper birds. In quite a few species, mostly of tropical or subtropical birds, the breeding pair is often assisted by a helper ready to feed and protect the young and to defend the territory from intruders. There are many possible explanations for such a behaviour: the helper is closely related, it acquires the know-how for raising young, it stands by for an eventual opportunity to take over. Maybe. But it could also be that the helper simply pays for the permission to stay by assisting the nest-owners, in the spirit of the *au pair* system.*

This example suggests that a reciprocating move can be addressed, not to the debtor, but to some of the debtor's relatives—here, to the nest-owner's young. In a sense, one might even say that the care expended by parents toward their offspring will be returned to their grandchildren. This may seem a bit far-fetched, but fits in with an intriguing proposal by R. A. Fisher: 'to regard the birth of a child as the loaning to him of a life, and the birth of his offspring as a subsequent repayment of the debt.'*

More obvious examples of reciprocation are provided by pairs of young male baboons. One of them picks a quarrel with the overlord of an oestrous female, while his comrade uses the diversion to mount her sneakily on the side. On the next occasion the two promising youngsters are apt to exchange their roles.* This is not quite the repeated Prisoner's Dilemma—the young baboons cannot get their reward in the same round—but it is a closely related game; again, reciprocation leads to cooperation. It reminds one of the trading between some Polynesian tribes who are forbidden, by a taboo, from ever meeting face to face. These tribes take turns in paddling to some appointed distant shore, where they leave their goods and pick up those left by their partners.

We may note that in this sort of iterated game, a defection usually brings the interaction to an end. In many applications, therefore, the probability of a repetition is not constant (as we have assumed so far), but subject to decisions by the players. It is clear that their readiness to continue is likely to depend on the history of the game so far. If it bodes ill for the future, they will tend to break off and look for other partners. It could be, of course, that there is no way out; this holds for couples of Roman Catholic faith. But many inter-actions can be called off, either unilaterally or by mutual consent. M. Feld-man and E. Thomas have studied such iterated Prisoner's Dilemma games in what they call the *behaviour-dependent context*: the probability of another interaction depends on the decisions of the players in the previous round.* Often, this makes it easier for *Tit For Tat* to invade a population of defectors: the games of a defector will mostly be short and therefore of very little value. But it may also happen that defectors can invade a *Tit For Tat*-population, using the time-honoured principle to 'Take the money and run'. In fact, this can even lead to a stable coexistence of *Tit For Tat* and all-out defection.

Clean choice

In Axelrod's tournaments, the players had no advance knowledge of their opponents' tricks and foibles; they were not allowed to watch the games of others from the sidelines. It is quite different with tournament chess players, who frequently build up a considerable store of knowledge about their opponents before ever meeting them. For the more down-to-earth inter-actions in other groups of primates, the character of each individual may like-wise be commonly known.

In a Prisoner's Dilemma game, a reputation for being a defector will be harmful; so will a reputation for being an all-out cooperator. But a reputation as a *Tit For Tat*-er is a tremendous bonus. It makes one sought after by players who wish to cooperate, and avoided by players who are up to no good. This is yet another way for *Tit For Tat*-players to form clusters, and to invade a population of defectors.

Preferential assortment helps cooperation to spread.* Most models assume random meeting (and, in genetics, random mating), because it simplifies the equations; but animals seem to pay no heed to this fact. Defectors will often be shunned by the community and enjoy less than their share of interactions.

Of course, neither reciprocity nor preferential assortment can work between anonymous individuals. The ability to recognize each other is there-fore essential. Most higher animals are able to identify their partners by sight, sound, or smell. In the human brain, selection has led to the specialization of a region devoted to facial recognition. If it is damaged, the traits of even the closest relatives can no longer be recognized, although vision remains essentially unimpaired. Simpler organisms use their sense of touch to make sure of their partner: in fact they *keep in touch*. Axelrod and Hamilton point out that most symbiotic relationships are based on permanent contact—for example between hermit crabs and sea anemones, or between bacteria living in the gut and their host organisms.*

A fixed meeting place also helps to guarantee the permanence of a relationship. There are these so-called *cleaner fish*, for example, mopping up the teeth of much larger fish. They coolly swim in and out of a mouth which seems just the right size to gulp them down in one go. Both fish get an advantage from their interaction. The smaller one finds morsels of food. The larger fish gets clean teeth. But it will be sorely tempted to finish the session with a light snack. However, this is just not done. According to Trivers, this restraint is so deeply ingrained that even when the large fish is attacked by a predator, it does not just swallow the cleaner and take off; it warns the cleaner to exist by partly closing its mouth, and only flees after the cleaner has left.* (I hope that it will never be shown that the larger fish increase their chance for a getaway by spitting a brightly coloured cleaner in the predator's path.) Anyway, Axelrod and Hamilton point out that the thriving trade of an aquatic cleaner seems to occur only among fish with fixed abode, some home range which is part of a coast or a reef. No similar grooming has ever been found among the freely moving fish roaming the open seas.

Territoriality helps to increase the readiness to cooperate, at least with one's neighbours. The mutual mistrust which poisons interactions between gypsies and local residents is caused by the small likelihood of continued contacts. The twin influence of mobility and anonymity works upon car drivers, too. A coarse kind of ruthlessness seems to settle upon some otherwise gentle souls, as soon as they let the clutch in.

It could well be that the disposition for a sedentary way of life, which overtakes most of us in riper years, is a trick of genetic predisposition leading to more cooperative behaviour. Axelrod and Hamilton speculate, by the way, that the reason why ants frequently profit from mutualistic interactions with other species, while bees do not, rests in their being much more likely to stay in one place. The less permanent a relationship appears, the more it is threatened by all-out exploitation. A businessman is in for some pretty painful experiences with his partners, as soon as rumours about his impending bankruptcy spread. We can point to a similar situation with some microbes living in our guts. They are completely harmless, and even beneficial, as long as circumstances remain normal; but they switch to a virulent state as soon as there are signs of injury, illness, or senescence of their human host.* These signs make it less plausible that the cooperation will endure. They portend the forthcoming death of the host organism, and in fact even hasten it by triggering the virulence of the microbes.

A free ride on the commons

This last example should actually be viewed as a Prisoner's Dilemma, not so much between microbe and host, as between microbe and microbe. The host, after all, has no leeway for decision; it simply constitutes the common resource of the microbes. It is in the microbes' interest to exploit it with restraint, since this guarantees a steady living. However, from the point of view of the individual microbe, it would be better still to restrain itself a little bit less than its fellow microbes do. If all microbes rush to this conclusion (metaphorically speaking, to be sure), this leads to unbridled exploitation and speedy ruin.

This variant of the Dilemma has become widely popular under the name *Tragedy of the commons*.* The principle of husbanding resources was known in the Middle Ages at least as well as it is known today, and villagers managed their own lands with restraint. But it rarely worked out for the commons, the piece of ground for collective use by all villagers. The reason is obvious. If some villager grazed an extra sheep belonging to him on the commons, the advantage was all on his side, while the disadvantage due to over-grazing was shared with the whole village. The only way that the neighbours could make up for it was by bringing some supernumerary sheep of their own to the commons. So the pasture was over-exploited by all, and soon ruined. Today, with a few exceptions, commons have disappeared. But the *Tragedy* stays with us. Right now the oceans are our commons.

Natural selection seems to be able, however, to program some parasites for restraint in exploiting their host. Up to the 1960s, this used to be explained by group selection. The parasite's restraint was supposed to have evolved because it helped the species to survive. But of course, if a more virulent strain could spread faster, then it would spread, even if this ultimately brought down the whole species.

Nevertheless, there are plenty of parasites whose virulence has decreased; some have actually evolved into mutualists. According to the evolutionary ecologist Paul Ewald, whether a parasitic species grows more or less virulent depends on how it is transmitted. If it needs the host for spreading further, its more benign strains will be favoured by selection; for virulent mutants will destroy their own carrier, and so ground themselves.* In this way, individual selection can effectively prevent the over-exploitation of a resource.

Could it work for us too? Could selection teach humankind to husband its environment? This, sadly, seems to be out of the question. We do not use planets for spreading further. So we seem left to face the *Tragedy* with nothing but our foresight.

The *Tragedy of the commons* is a Prisoner's Dilemma with more than two players. Society has a similar problem with so-called 'free riders', who use a common institution (like a town's bus system) without paying for it. What seems to hold in all such interactions is that the larger the group, the harder it becomes to achieve cooperation. If one punishes a defector by defecting in turn, one punishes the whole community, including the innocents. And even if there were a way to aim the retaliation exclusively against the miscreants, there remains the problem of who is going to bear the costs of punishing them. The community, in all fairness, of course. But this offers a new opportunity to defect: by not participating in reprisals and letting others do the police job.* Therefore, it is not just the defector who should be punished, but everyone refusing to punish the defector; and, yes, everyone refusing to punish those who refuse to punish. Et cetera. Infinite regress. It seems difficult to overcome this problem without a central authority. And who should police *them*? It is a familiar old tune. The Dilemma keeps us its Prisoners.

9

Playback

Hidden within any true man, there is a child that wants to play,
says Nietzsche.—Why hidden?, asks my wife.
Konrad Lorenz

Horseplay

Playfulness signals a lack of maturity. Now, it seems that the delaying of maturity played a significant part in human evolution, and accounts for our so-called *neotenic* properties:* bodily signs of childishness, like the high forehead, the delicate chin, the position of the first toe, or the sparse growth of body hair. Humans are, to put it in a nutshell, immature apes. We need one-third of our lifespan for growing up, and we become adults at an age when most primates are already dead. No other animal remains dependent for as long as we do. This gives us plenty of time for learning, but especially for playing. We *gamble away* the time allotted as lifespan to our nearest relatives in the animal kingdom. Luckily, we are granted an extension of the game.

We humans have, of course, no monopoly on play. Playful behaviour can be observed in many birds and mammals. A young horse romping about quite obviously *enjoys* itself. The Viennese *fin de siècle* author Peter Altenberg wrote of a little girl who wished to be transformed into *two* small dogs playing together. The pleasure in having a reliable companion for play was certainly an influential factor in the domestication of some animal races. Which reminds one that Konrad Lorenz (who, incidentally, wrote at length on the extended period of childhood in humans) viewed the emergence of humankind as an instance of self-domestication.* Did we possibly breed ourselves into playmates? This could have worked by positive feedback, like so many other extravagances of selection. If peacock females can breed for the long tails of peacock males, why should not humans breed for a strong play-urge in their children, by caring more for those who are fun to play with?

In any case, play is (in contrast to laughter) not uniquely reserved for humanity. It is widespread among higher animals.* Birds glide in the storm, monkeys cavort in the trees, and colts gallop in the fields. River otters take turns in sliding down muddy river banks. Lion cubs spar and tussle. Primates play peekaboo. 'When I play with my cat', wrote Montaigne, 'who knows if she doesn't have a better time than I do?' Being in the mood for playing is indicated by its own set of signals: a young coyote with upright ears, round eyes, and a slightly open mouth is ready for some frolicking. In most cases, just watching a game triggers the wish to join in. Playing can be more infectious than yawning.

Play is a goal in itself, not a means to an end. It is never of immediate use, but often of value in the long term. It has no purpose, but it makes sense. A good part of it seems to be innate. Thus young seals exhibit elements of play which are typical for terrestrial predators. The possibility that the seals imitate grown-ups can be excluded. The playing behaviour of seals must have been transmitted genetically over millions of years.*

In most cases, a young animal at play is developing activities which it will need later on, for catching prey, for taking flight, or for courting and wooing. When a cat shakes a rag, when it swishes with the paw beneath a carpet, or when it jumps after the end of a string, it obviously displays its repertoire of hunting behaviour. Scuffling whelps gain, usually without hurting each other, rich experience for later, more serious fights. In their pursuit games, they are apt to exchange roles at a moment's notice, from hunter to hunted and back again. What are they feeling in such a moment? The cat which approaches stealthily, all trembling with excitement, and suddenly jumps upon one of my discarded socks, is certainly not the victim of an error. But it is 'deluded' in the original sense of the word (which, as mentioned already, contains the Latin for 'play'). The cat *pretends* that the sock is a prey. It obviously does it for fun, and does not need the reward of a piece of smoked salmon, say, to be ready to repeat the exercise. The play is its own reward. Such skill-forming types of play are enjoyed for the same reason as eating or copulating: because they are advantageous in the struggle for life. Among rhesus monkeys, for instance, adolescent males play almost three times as much as females, because they need all the training they can get for the next mating season.

Young animals playing together may sometimes form friendships which can last for life. And they can also gain useful experience in the playful exploration of their community and neighbourhood. Inquisitive behaviour usually has a childish, playful streak. Curiosity can sometimes be costly— 'curiosity killed the cat'—and play is liable to be dangerous, more dangerous, at any rate, than doing nothing. But the gain in experience and skill seems to be worth the risk. It comes as no surprise that the most playful animal species are intelligent, live in groups, and lead a wide-ranging life: wolves, for instance, or hyenas, lions, river otters, bottle-nosed dolphins, orcas, parrots, elephants, polar bears, brown bears, chimpanzees, and humans.*

Even the most developed play-urge appears to be weaker, however, than other instincts. The merest hint of fear or hunger is enough to bring play to an end. Play can only thrive if not nagged by necessity. It is stifled by any form of external constraint. This reminds one of the concept of *play* used for mechanical parts as 'limited mobility'. Closely interwoven in the idea of play are the three notions of freedom, motion, and a space reserved for it. We have seen that the *free play* of a biological molecule is essential for its evolutionary development. The looser the constraints, the richer the possibilities which can be explored.

In this sense, even *innate* playing behaviour is free from constraints. It is preprogrammed openness. This explains why it withdraws discreetly as soon as other instincts come to the fore. Games need leisure: they are programs for time off. This openness makes them more difficult to describe than any other biological function—Humankind knows a considerable number of sexual positions; the kitchens of the diverse nations exhibit a lot of variety; but all this is as nothing, compared to the diversity of games. Psychologists have made attempts at classifying them: they speak of function games, fiction games, construction games, and so on. But it is not always simple to find their proper label.

Language games

The Kunsthistorische Museum of Vienna boasts a Breughel depicting several hundred children's games. Many of them are no longer seen nowadays, but lots of others have come along in their stead. A brief look into a catalogue of computer games shows that their manufacturers are not idling their time away. Or all those board games—most of us may be acquainted with half a dozen or so, but every year, many novelties appear on the market. Who is devising them all? They fill the shelves of game departments right to the top. And then all those card games. Puzzles. Games of patience and of skill; toy bricks; meccanos and legos; model trains; dolls and teddy bears; spin-tops and rattles. We have reached the baby corner. Is a comforter still a toy? Is thumb sucking a type of play? Our way back leads through the sports department. Roller-skates, tricycles, sledges. Ball games. Why do we enjoy such an elementary pleasure in throwing, hurling, or kicking a ball? Team games, combat games, pursuit games, games of motion and agility. Indoor games and outdoor games.

What is common to all games? Wittgenstein pondered this question in his *Philosophical investigations*.* Is there always an element of competition present? 'Think of patience.' Must there be losers and winners? 'If a child throws a ball against a wall and catches it again, this feature has disappeared.' 'Look for the role played by skill and luck.' 'And now think of games like ring-a-ring-a-roses.' What is still a game and what is no longer? 'Family resemblances show up and vanish. One does find correlations, but nothing common to all.' 'Can you trace a frontier? No.'

Wittgenstein uses 'game' as an *example* of a notion with blurred borders. (The German word for 'example', incidentally, is *Beispiel*, which contains -*spiel*, meaning toy or game.) It does not matter that the rules for handling such a word are not sharply defined, as long as they suffice for its role in the *language game*. If one wants to understand the meaning of a word, one has to ask oneself how one has learned to use it: 'by means of what examples? In which language games?'—Wittgenstein mentions other fuzzy words, for instance 'number', or 'good', or 'to know'. But he obviously prefers the word 'play', possibly because playing is the most open of all activities. In fact, his question 'How is the concept of a game limited?' verges on sophistry. Small wonder that the *notion* of a game has vague limits, if playing *itself* needs fuzzy borders.

Openness, freedom from biological necessity, was essential for the emergence of culture. In his *Homo Ludens*, the Dutch historian Huizinga presented what has become a classic analysis of play in civilization.* He showed that there is hardly any domain of human culture which has not been deeply affected by play. The rules of games developed into the self-imposed rules of civilized communities, for instance in the ceremonies and festivities which were of such enormous importance for early societies. Festive play is a major part of every cult. The spoilsport is an outcast (far more than the cheater, by the way).

Ritualized play-rules are also essential in law. Even today, every lawsuit contains streaks of mummery and play-acting, as well as elements of tournament games and games of chance. Tension and uncertainty of the outcome always belong to the game. We can find this again in the spheres of trade and industry. The word 'gain' has a connotation of gambling (which 'salary' has not). Every speculation is a game of chance; the ratings at the stock exchange describe random walks.* And a driving element in economy is contest, of course: the eternal desire to be (and do) better than the other. In trading, this urge to excel can frequently be counterproductive, as we have seen with the Prisoner's Dilemma.

It is in the domain of contests and fights that the relationship of cultural with biological games is particularly close. The rules of combat sports display amazing parallels to the ritualized fighting of animals. One species of cichlids has been christened 'Jack Dempsey' for its fairness and pugnacity.* Hunting, too, is thoroughly spiced with elements of play, even (and especially) in archaic societies where it is a vital necessity. This is still echoed in the English word 'game'. And a fox hunt is a game—as cruelly one-sided as a cat playing with a mouse. Even war is intimately associated with play: this is shown by expressions like 'war game' (or *Kriegsspiel*, in preparation of the so-called *Ernstfall*), or by staff exercises in *sand-boxes*. Chess is purported to be a distant cousin of such martial thought experiments, and the mathematical theory of games once held great appeal for military planners.

It is not far from war to love. The ritualized aspects of both fighting and wooing are side-effects of sexual selection. This is strikingly illustrated whenever a specimen of the Siamese fighting fish meets its like: it is only through the first stages of their display that the two fish learn about the sex of the other, and proceed in a smooth transition to either competition or exhibition.*

In sexual encounters, games are mainly used for the introductory phase: flirting, dallying, wooing, the pleasant detours of the *foreplay*. Even some children's games, like 'playing doctor', have a sexual component. When a priest in the confessional asks a boy whether he has 'played with himself', he usually refers to a sexual thought experiment. Such thought experiments are covered (if this is the appropriate expression) by *Playboy* magazine. Its Latin edition would be called *Homo Ludens*, one assumes. And what is a Playmate, then, but a model?

The small step from display to performance leads us to art: comedy plays, tragedy plays. Acting is so close to the role-games of children that one is led to suppose that performers came first and their public only much later. In other, non-dramatic fields of literature, the play character comes from playing with words. Or is it mostly a play with emotions? We are moved. Play is movement. In movies and dances, games of motion became art forms.

The connection of music with games is apparent in expressions like the French *prélude* or the German *Spieluhr*. And musicians are said to 'play' their instruments, in English, French, German, Slavic, and Arabic. As for the plastic arts, experience shows that children and apes do not hesitate to use colours and bricks for play. And is painting not the art of illusion (the word comes from *in-ludere*), of *trompe-l'oeil*, of make believe? Turning now to architecture, we may deem it at first as too tightly related to the necessities of life to be viewed as a game: but even a functionalist like Le Corbusier defined it as 'the scientific, correct and magnificent *play* of volumes assembled under the light'.* The pyramid builders could not have put it more neatly.

From pyramids via sphinxes to riddles, and we have reached science. Many legends and myths bear witness to the importance of question-and-answer games, mostly in the form of a riddle which decides on life or death. (In German, the same verb *lösen* is used for solving a riddle and for untying knots.) These riddles were often mere jests or plays with words, but sometimes they anticipated basic problems of science which any child is likely to be curious about, like 'Where do we come from', etc.

Even today, the rituals of academe still contain elements of play: examinations as question-and-answer games, for instance, or the public disputes, or the sectarian consecrations, promotions, and inaugurations. The word 'school', incidentally, stems from *schole*, which in ancient Greek meant leisure or play-time. For today's school-children, this may not seem very funny.

From model to math

Even those who feel that play is sadly neglected in today's education must admit that contemporary research still has an intimate relationship with playful exploration. This is shown, among other things, by the growing tendency to speak of *models* rather than theories. Like toys, such models can be displayed, dissected, and (often) discarded. Nobody would be prepared to go to the stake for their sake. Models have a kind of intermediate reality. What could be more fleeting than so-called *quantum reality**—it is but a shadow play, as in Plato's cave myth. On the other hand, models lend themselves very well to being handled and tinkered with. Tinker Toys were used in the discovery of the double helix.

Another point shared by science and play is that both, in most cases, are not intended for direct profit. Scientific investigations which promise immediate applications are easier to fund, admittedly—today's trend towards project-oriented research makes this very clear—but they are not necessarily of lasting value. Astrologers and alchemists, too, aimed for extremely useful applications, but this was precisely why they failed so abjectly. 'As long as alchemists looked only for the philosopher's stone', wrote Boltzmann, 'all their attempts were fruitless; only the limitation to seemingly less important questions created chemistry.'*

Mathematics is surely the most playful of all sciences, although its entertainment value is sometimes overlooked. It is impossible to say whether *Life*, for instance, is more mathematics or game. The mathematical axioms correspond to the rules of a game. Both axioms and play-rules create a space of their own, a playground, a world within the world. Children at play can become so excited that they get *beside themselves* (whereas mathematicians traditionally become *absent-minded*). Neither play-rules nor sets of axioms can be reasonably held in doubt; they can only be accepted or not.

Axioms may be viewed as specific strings of symbols which can be transformed, according to precise rules, into other strings of symbols—the theorems (which correspond, for instance, to all reachable positions at chess). This, at least, is the mathematicians' newspeak for official use;* in fact, it rests on a confusion of form with content. A formal concatenation of symbols has not much more to do with mathematics than it has with music or poetry (which could be said, with the same justification, to consist in correctly formed sequences of letters or notes). Those strings of symbols require translation—for music into sounds, for mathematics into concepts—to acquire their meaning. (It is for *proving* things about mathematics that the 'strings-beget-strings' view is useful.) Only the associated images yield a sense to mathematics. *Imagination* is the most important tool of mathematicians. It is also the most important requisite for child's play. The derivation of correctly constructed formulae can be done nowadays by computer. But computers cannot replace imagination. They can be priceless, of course, in stimulating it.

To do *as if*, to pretend, is an essential tool in the mathematician's trade. This is particularly well exemplified in *indirect proofs*, where one pretends that this or that statement is valid and derives implications which in the end contradict it. Some contradictions of so-called common sense, on the other hand, are based on false appearances only. The play with paradoxes tickles the imagination—'What is as large as part of itself?' could be a riddle from a fairy tale, but in set theory, it leads to a workable definition of infinity. The paradox of the liar, or that of Achilles and the tortoise, played important roles in the development of logic and infinitesimal calculus. The theory of probability originated, as we have seen, in paradoxes from gambling. And the game of squaring the circle is as fascinating for today's mathematicians as it was more than two thousand years ago for their Alexandrian brothers-in-trade. It keeps offering new and surprising aspects. Only recently it was shown, in fact, that a circular disk can be dissected into a finite number of pieces which, if properly reassembled, form a square.* (Don't try too hard to visualize it!)

Such circle-squaring is obviously a result without any possible use. And yet—who knows? The history of mathematics is not without unexpected applications. In *A mathematician's apology*, Hardy* writes that whatever the drawbacks of higher mathematics, nobody could accuse it of helping to make war. As examples of fields which could certainly never be brought to military applications, he mentioned the theory of prime numbers and the theory of relativity. This was in 1940. In the meantime, the armed forces of many nations routinely use for encoding their messages so-called *public-key* methods which are based on results from prime number factorization;* and, yes, the theory of relativity also appears to have stumbled upon some strategic applications.

This does not mean, of course, that all mathematics will ultimately find its use. Neither does it imply that mathematics is justified by the vague chance of possibly finding an application in some distant future. A game does not need any *justification*.

Endgame

Comparing mathematics with chess is particularly popular. Not so much with tournament chess (although mathematicians are a competitive breed), but with the solving of chess problems. Can Black checkmate White in three moves? Problems like this are found in the games' corners of Sunday papers, well divorced from all their other contents. A clear *demarcation* is a frequent prerequisite for a game. The game is open, but not unlimited. The rules of the game define a space, be it a field, a stage, a ring, a ceremonial enclave, or a court of justice. In antiquity, even battle fields were sometimes staked out in advance. The game board reflects, or better, models this. Similarly, mathematical *spaces* are ruled by their own agreed sets of laws. To some, they may offer an opportunity to withdraw from the outside world.

But there are only a few mathematicians who really do believe that they are playing. Some used to say so because it is a simple and incontrovertible formula for silencing obnoxious philosophers, for instance. Unfortunately, these philosophers are apt to take the answer at face value. Another inconvenience is that it can make the mathematicians forget to account to themselves for what they are doing; and this, in turn, can effectively lead to mere toying. The excuse becomes a self-fulfilling prophecy.

There exists no test for deciding when mathematics is of value, but a large part of good mathematics is *motivated* rather than *applied*.* In this respect, mathematics owes a great deal to biology (though not as much as to physics, certainly). Some of the most interesting developments in cellular automata, game theory, dynamical systems, statistics, and probability theory have been due to biological questions. Biological models are frequently more accessible than physical models; they are also, it must be said, less demanding in general. It is like the difference between pop and classical music. Biomathematics has not yet provided mathematical symphonies as glorious as the theory of relativity or statistical mechanics, but it has given us some catchy little tunes.

But biologists will not content themselves, naturally, with being praised for motivating mathematicians. In the spirit of reciprocal altruism, they feel entitled to a return; some, in fact, feel that they have been cheated of it. The current models for artificial life, for instance, enjoy a far more enthusiastic reception from computer freaks than from biologists. Similarly with neural networks: most mathematical models, there, look like nothing seen in nature. Models for population growth, too, are brushed aside by some ecologists as being far too simple to be real. But the models are intended as thought experiments, and not as pictures of real life. A model must be properly reduced, to the point of caricature, for the decisive elements to stand out. Such a model promises no prediction, but it offers understanding. Abstraction is always caricature: an exaggeration of the essential traits, or at least an omission of irrelevant detail.

We know by now that some of the simplest models can act in the most unpredictable ways; in a sense, this is the *moral* of the story. Those who have always known and said that life is uncomputable will draw solace from it. We are still, like Newton's children, *playing* on the shores of an ocean. Thought experiments and mathematical models can sometimes yield insights into the limitless complexity created by billions of years of natural selection, but it is probably hopeless to expect detailed predictions.

Hopeless but not serious, as the saying goes. For, to quote Sigmund Freud:* 'The opposite of play is not seriousness, but—*reality*.'

Endnotes

Chapter 1: notes and references

1 *The monk with the peas*: for example, see Orel (1984).
1 *A young mathematician*: Fisher (1958), p. viii of the preface.
2 *Self-ref and self-rep*: Hofstadter (1980), in particular Chapter 26.
3 *An ever increasing role in biology*: see, for instance, Murray (1989).
4 *Thought experiments*: Sorensen (1991).
5 *The ultimate in technology transfer*: Stewart (1989), p. 291.
5 *Evolution is tinkering*: Jacob (1982).
5 *Strange developments*: see Riedl (1979).
6 *Differences in training*: Fisher (1958), preface.
7 *Illustrations of the highest doctrines*: Maxwell, *Introductory Lectures on Experimental Physics*.

Chapter 2: notes and references

The game of *Life* straight from the horse's mouth, so to speak, can be found in Chapter 25 of Berlekamp *et al.* (1982). A more detailed, brilliantly written account of *Life* and Conway's self-replicating automaton is given by Poundstone (1987). The two volumes edited by Langton (1989) and Langton *et al.* (1991) contain a wealth of material on artificial life.

9 *Impeccable techniques of mathematical logic*: Dowling (1990).
9 *Major public concern*: Peter Denning keeps a running watch on computer crimes in *American Scientist* (see, for example, **76**, 236–8 (1988); **77**, 126–8 (1989); **78**, 10–12 (1990)).
10 *Legendary column*: see Gardner (1970). Many follow-up articles appeared in Martin Gardner's column in *Scientific American*, as well as in the last three chapters of his book (Gardner 1983). The tradition was carried on by A. K. Dewdney in his column *Computer recreations* and in his book (Dewdney 1989).
10 *Life was something special*: Berlekamp *et al.* (1982).
12 *Life is a spectator sport*: see Poundstone (1987).
13 *An example of a cellular automaton*: the bible on cellular automata is Wolfram (1986).
14 *Looking for glider guns*: see Berlekamp *et al.* (1982) and Poundstone (1987).
15 *Artificial life is good copy*: see the two books edited by Langton (1989, 1991), which contain a wealth of material on the sources of artificial life.
16 *and to Princeton*: see Heims (1980) and Legendi and Szentivanyi (1983); Poundstone (1987) also contains biographical material on John von Neumann.

19 *Tiny homunculi*: for a description of the 'spermist's' views, see Jacob (1976).
20 *Alan Turing*: see the fascinating biography by Hodges (1983).
20 *The universal computer simply mimics*: Turing (1936–7). For a good introduction, see Dewdney (1989), Chapters 28, 48, 55.
20 *Second-order machines*: see the superb introductory essay in Langton (1989).
21 *Universality and self-reproduction*: see Dewdney (1989).
22 *Stanislas Ulam*: see Ulam (1976).
22 *Experience with cellular automata*: Ulam (1970).
22 *Did not live to finish his book*: John von Neumann's book *Theory of self-reproducing automata* was completed and edited by A. Burks in 1966. A companion volume edited by Burks (1970) contains important papers extending John von Neumann's work.
23 *Codd's cellular automaton*: Codd (1968); see also Chapter 43 in Dewdney (1989).
23 *A still simpler example*: Langton (1984).
24 *Automata which were still smaller*: Byl (1989).
24 *Genetic take-over*: Cairns-Smith (1982). See also Chapter 6 in Dawkins (1986).
26 *Examples of life themselves*: Langton (1986).
26 *Partial self-repair*: Laing (1977).
26 *Lamarckian evolution*: see, for example, Chapter 11 of Dawkins (1986) or Chapter 1 of Maynard Smith (1989).
26 *Telegram to China*: the quote is in Maynard Smith (1989), p. 9. For more, see Weismann (1904).
27 *Drive for self-improvement*: see, for example, Chapter 9 of Dawkins (1982) or Gould (1979).
27 *Collision of 13 Gliders*: see Berlekamp *et al.* or Poundstone (1987).
37 *Fermat's conjecture*: see Stewart (1987), Chapter 3.
37 *Whether it will ever halt*: see Stewart (1987), Chapter 19.

Chapter 3: notes and references

For a general introduction to theoretical ecology, see Real and Brown (1991) and, for a more mathematically oriented set of classical papers by Volterra and others, Scudo and Ziegler (1978). For a collection of up-to-date surveys, see Roughgarden *et al.* (1989). The recent book by Renshaw (1991) gives a good account of population dynamics. For excellent introductions to chaos, see Gleick (1987), Ekeland (1988), and Stewart (1989). For a mathematical introduction, see Devaney (1986).
41 *Shark and fish*: Dewdney (1984).
43 *Senator Vito Volterra*: see Chapter 5 in Kingsland (1985) and Scudo (1970).
44 *Failures in insect pest control*: see, for example, Roughgarden (1979).
44 *The war game imitates the armed conflict*: A. Lotka, see Kingsland (1985), p. 109.
44 *Dry as dust*: Hamilton (1991).
44 *Three brilliant Russians*: see, for example, Scudo and Ziegler (1978).
45 *Cycle as regularly as an alternating current*: see, for example, Hofbauer and Sigmund (1988).
45 *Paradox of enrichment*: see, for example, Roughgarden (1979), p. 447.

45 *Some ecologists quibble*: see, for example, Moran (1953), Gilpin (1973), Schaffer (1985), and Renshaw (1991).
46 *Scavengers without the patience*: Colinvaux (1978).
46 *In a review of eight pairs*: Tanner (1975). For a recent study showing that most predators have little impact on their prey, see Paine (1992).
46 *Field studies in Newfoundland*: Bergerud (1983).
47 *In a great many free-living populations*: see May (1976), Schaffer (1985), Renshaw (1991).
48 *Flies kept in jars*: Nicholson (1954).
49 *Wittgenstein quote*: Wittgenstein (1953).
49 *With rather blurred vision*: MacArthur (1966), p. 156.
49 *No standing room for the progeny*: Darwin (1859), p. 64.
50 *Water fleas in a jar*: Pratt (1943). See also Renshaw (1991).
52 *They did their best to hide it*: Hamilton (1991).
52 *Simple models with complicated dynamics*: May (1976).
53 *Eclipses of the moon*: Ekeland (1988).
54 *The flap of a butterfly's wing*: Gleick (1987).
55 *The story of Mitchell Feigenbaum*: see, for example, Stewart (1989); Ekeland (1988); and Gleick (1987).
55 *Spilled into many other fields*: see, for example, Hall (1992).
56 *It is quite credible*: Darwin (1859), p. 74.
56 *Throw up a handful of feathers*: Darwin (1859), p. 75.
56 *Three possible scenarios*: see, for example, Hofbauer and Sigmund (1988).
57 *Done by G. F. Gause*: see, for example, Colinvaux (1978).
58 *The notion of the niche*: see MacArthur (1958); Hutchinson (1965); Kingsland (1985).
58 *And even of three lab populations*: see Weissing (1991).
58 *Possible worlds*: Haldane (1927).
58 *Two radically different results can emerge*: May and Leonard (1975).
59 *Systems with 10 links have been observed*: Cohen (1989).
60 *Three-layered feedback devices*: Hastings and Powell (1991); Jansen and Sabelis (1992); Muratori and Rinaldi (1992).
60 *From an intertidal community*: Paine (1966).
61 *Population sizes may spin chaotically*: Gilpin (1979).
61 *Not enough to mediate a permanent co-existence*: Hutson and Vickers (1983); Kirlinger (1986).
61 *Such a switching predator*: Hutson (1984).
61 *Newfoundland's ecology*: Bergerud (1983).
62 *Lotka's very first model*: Lotka (1923).
62 *If not one head of game*: Darwin (1859), p. 68.
62 *Parasites can drive oscillations*: see the monumental survey by Anderson and May (1991). For an example of a virus of a virus of a virus, see Nee and Maynard Smith (1990).
62 *A cybernetic universe*: Ray (1991).
62 *The most sophisticated artificial life*: Rennie (1992).
63 *Under pressure to treat it well*: Ewald (1987).
63 *The patchiness of the environment*: see, for example, Huffaker (1958); Hassell *et al.* (1991); and Renshaw (1991).
63 *The orderly succession*: see Colinvaux (1978), Chapter 12.

64 *Fumigated a few mangrove islands*: Simberloff and Wilson (1969).
64 *The worst ecological catastrophe*: Gould (1983), p. 347.
64 *The count was again 25 families*: May (1978).
64 *In environments of structure B*: MacArthur (1972).

Chapter 4: notes and references

A highly successful introduction to probability theory is the classic textbook by Feller (1968). The neutral theory of molecular evolution is superbly expounded in Kimura (1983).

67 *sooner or later*: for a mathematical and historical introduction to branching processes, see Jagers (1975). Let us briefly sketch the flaw in the original argument. We denote by $f_0, f_1, f_2, \ldots$ the probabilities that a man has no son, or one, or two, etc. Since one of these events is bound to occur, these probabilities sum up to 1. The event that the line issued from a given individual A vanishes (which has probability q) can be split into disjoint events as follows: either A has no son at all (probability f_0), or A has one son (probability f_1) whose line goes extinct (probability q), or A has two sons (probability f_2) whose lines both go extinct (probability q^2), etc. This yields

$$q = f_0 + f_1 q + f_2 q^2 + \ldots$$

Obviously $q = 1$ is a solution to this equation. What was originally overlooked was the possibility that there might be other, smaller solutions to this equation.

67 *A man has with probability x so many sons*: Lotka (1931).
69 *A closer look*: see, for example, Karlin (1966); Jagers (1975).
72 *Forty times more DNA*: see Maynard Smith (1989), p. 204.
73 *Allometric limitations*: see, for example, Gould (1982), Chapter 21.
73 *Eigen's theory on the origin of life*: see Eigen (1971); Eigen and Schuster (1979); and the expository paper by Eigen *et al.* (1982).
73 *Sceptical on-lookers*: see, for example, Shapiro (1988).
74 *Probability theory teems with paradoxical results*: for more, see the delightful book by Szekely (1986).
74 *Random walks*: see, for instance, Karlin (1966) and Feller (1968).
77 *The Ellsberg paradox*: for this and related material on chance and uncertainty, see Ekeland (1991).
77 *Positive feedback in business and biology*: Cohen (1976) and Arthur (1990).
79 *Flattering expressions*: all these quotations can be found in Szekely (1986).
79 *The common root*: from E. Schrödinger's inaugural lecture in Zürich in 1922: 'What is a law of nature?'
79 *Foundation of the edifice*: Monod (1972).
80 *Responsible for having met extinction*: for an introduction to this subject, see Gould (1983), Chapter 27. For more, see Raup and Sepkovski (1982).
81 *Which belongs to the classical repertoire*: see, for example, Crow and Kimura (1970).
81 *For amusement*: C. Darwin, autobiography (F. Darwin, 1887).
83 *It changes by less than 1 per cent*: Schuster and Sigmund (1988).
83 *More likely to occur in a large population*: Fisher (1958), p. 84.

84 *Twice as high as the selective advantage*: Haldane (1932), p. 114, and Fisher (1958), p. 84.

84 *Has been eliminated after seven generations*: Kimura and Ohta (1971), p. 3.

85 *Substitution rate equals mutation rate*: see Kimura (1983). For a short introduction, see Kimura (1979).

86 *Geneticist Sewall Wright*: see the biography of S. Wright by Provine (1986).

86 *Evolution and the genetics of populations*: Wright (1968, 1969, 1977, 1978).

87 *Fisher's recipe*: see Provine (1971).

88 *At the molecular level*: see Kimura (1983), Chapter 10.

88 *Hoarded over millions of years*: Dawkins (1986).

89 *Critics of the neutral theory*: see, for example, Gillespie (1991).

92 *The so-called junk-DNA*: Kimura (1983), p. 104.

92 *A common ancestor to us all*: the first report on Eve was issued by Cann *et al.* (1987). For a concise introduction, see Maynard Smith (1989).

93 *In a somewhat over-sanguine way*: see Templeton (1992).

Chapter 5: notes and references

The pioneers of population genetics have found fine biographers. We mention *R. A. Fisher: the life of a scientist* by Box (the scientist's daughter) (1978); *JBS: the life and work of J. B. S. Haldane* by Clark (1969); and *Sewall Wright and Evolutionary Biology* by Provine (1986).

Good introductions to the problems of evolution and heredity are Ridley (1985) and Edey and Johanson (1990). Some of the best modern texts on population genetics are Crow and Kimura (1970), Cavalli-Sforza and Bodmer (1971), Crow (1986), Hartl (1980).

95 *Maupertuis anticipated Mendel*: on Maupertuis and the Ruhe family, see Jacob (1976).

96 *Fisher sketched this approach*: Fisher (1958), p. 7.

97 *Mendel devised more feasible experiments*: see the monumental biography on Mendel by Iltis (1932); for a concise text, refer to Orel (1984).

97 *A sharp and certain separation*: Mendel (1865).

98 *This uncanny fit*: Fisher (1936).

99 *All diversity would cancel*: this argument was first used by Darwin's contemporary Fleeming Jenkin; see Dawkins (1986), p. 113.

99 *Tantalizingly close*: see Hartl (1980), p. 63.

100 *Let theory guide your observations*: see Gruber and Barrett (1974).

100 *An extra sense*: DeBeer (1964).

100 *And then turned to discussing the* Origin: see Hartl (1980), p. 69.

100 *Hardly reconcilable at first*: for a highly readable account of the initial misunderstandings between Darwinists and Mendelians, see Provine (1971).

101 *How utterly simple*: see E. Mayr's introduction of the 1964 re-edition of Darwin (1859).

101 *RNA-molecules in a test tube*: Eigen *et al.* (1981).

101 *Consist of identical copies*: well, not quite. For more precisions, see Buss (1987).

101 *Some 1.5 billion years ago*: this date is actually far from certain. See, for example, Jacquard (1990).

103 *Seven loci sit on four chromosomes only*: see Hartl (1980), p. 13.

104 *the carriers of heredity*: see Jacob (1976).
104 *During a cricket match*: see Provine (1971).
104 *More than elementary*: Hardy (1908).
105 *Little-known German physician*: Weinberg (1908).
105 *America would pale*: Jacquard (1974), p. 96.
106 *Freed from so-called impurities*: see Müller-Hill (1988) and Proctor (1988).
106 *During some 60 generations*: see Falconer (1981), p. 22.
106 *Safety margin*: see Sved and Mayo (1970).
107 *Another locus may decide*: Fisher (1958), p. 52.
107 *Dominance modifiers are quite common*: see Sved and Mayo (1970).
107 *Assisted a winner in spreading*: Wagner and Bürger (1985).
107 *Another road to dominance*: Sheppard (1958), Chapter 8.
108 *The Fundamental Theorem of Natural Selection*: Fisher (1958), p. 22.
108 *My own view*: Maynard Smith (1989), p. 118.
108 *Population would cycle*: Koth and Kemler (1986).
108 *Thorny thicket of alternatives*: Karlin and Feldman (1970).
109 *Curious habit of conjugation*: Jacquard (1990), p. 37.
109 *Tested against a maximal diversity of background*: Crow (1979).
111 *A defence against outlaw genes*: Haig and Grafen (1992).
111 *Incest deserves its bad reputation*: see, for example, Falconer (1981).
112 *Degree of relatedness*: see, for example, Crow and Kimura (1970).
112 *Which prevents mating between close relatives*: see, for example, Krebs and Davies (1987), p. 259.
113 *Probe the incoming pollen*: Karlin and Lessard (1986), p. 245.
113 *The first to explain this was R. A. Fisher*: see Crow (1986), p. 41.
114 *Go on for perpetuity*: Darwin (1859), p. 101.
116 *It takes some computation*: Bartz (1979).
117 *Living under bark*: Hamilton (1978).
117 *Most remarkable discovery*: Sherman *et al.* (1991).
118 *Some of the most brilliant pages in the* Origin: Darwin's treatment of sterile castes is in Darwin (1859), pp. 235–42.
118 *As Alexander emphasizes*: see Alexander *et al.* (1991).
119 *Bright warning colours*: Fisher (1958), p. 177.
119 *Three-quarters rule*: Hamilton (1964); Dawkins (1989).
120 *Correlations agree with Mendelian rules*: Fisher (1918).
120 *For instance in a breeding program*: Falconer (1981); Barton and Turelli (1989).
121 *It does not seem to matter which allele*: Wagner (1990).
121 *A bag full of coloured beans*: Mayr (1963).
121 *In defense of beanbag genetics*: Haldane (1964).
121 *Parental imprinting*: Rodgers (1991).
122 *Tempered by sloppiness*: Dyson (1988), p. 95.

Chapter 6: notes and references

Among the books devoted to sexual replication and its attendant riddles, we mention Williams (1975), Maynard Smith (1978), and Bell (1982). Many distinguished authors, including Hamilton, Maynard Smith, Hoekstra, Bell, Crow, and Williams have contributed to the volumes edited by Stearns (1987) and Michod and Levin (1988).

125 *Sexual Selection*: Darwin (1859), p. 87.
126 *Runaway process*: see Fisher (1958), p. 152.
126 *Antidote to stiff reading*: Bennett (1983), p. 20.
126 *Female ladybirds*: Majerus *et al.* (1983).
127 *And the test works*: Partridge (1983).
127 *Moves from the parasite*: Hamilton and Zuk (1982).
128 *Blood parasites*: see Trivers (1985). But some of the evidence does not look so good, see for example, Balmford and Read (1991).
128 *A bright red belly*: Milinski and Bakker (1990).
128 *Two sorts of feedback*: see Dawkins (1986), Chapter 8.
131 *It recognizes itself*: Dawkins (1986), p. 206.
131 *More sophisticated models*: see Lande (1981); Kirkpatrick (1982); O'Donald (1982); and Pomianowski *et al.* (1991).
132 *Females wear brighter colours*: see, for example, Trivers (1985), Chapters 9 and 10.
132 *Glorious exhibitionism*: Huizinga (1970).
132 *Copulate with blobs of wax*: Maynard Smith (1988).
133 *By positive feedback*: Parker *et al.* (1972). See also Charlesworth (1978).
135 *A negative feedback*: Fisher (1958), p. 158.
135 *Three months after conception*: Trivers (1985), p. 303.
136 *Batches of silversides*: Conover and Voorhees (1990).
138 *The mealy bug*: Trivers (1985), p. 286.
138 *Equal investment*: see, for example, Maynard Smith (1978) and Charnov (1982).
139 *Extraordinary sex ratios*: Hamilton (1967).
139 *Whichever sex is left out*: Maynard Smith (1988).
139 *Incompatibility types*: see, for example, Karlin and Lessard (1986).
139 *Underequipped cells*: Parker *et al.* (1972).
140 *Explanation for having two sexes*: Hurst (1992).
141 *According to Hurst and Hamilton*: Hurst and Hamilton (1992).
141 *Evolution of sexes*: Hoekstra (1987) and Hoekstra *et al.* (1991).
141 *Why in fact always two?*: Fisher (1958), p. ix.
141 *Based on a paradox*: see, for example, Maynard Smith (1988).
143 *Twofold cost of sex*: Maynard Smith (1971).
143 *A laconic hint*: Fisher (1958), p. 136.
143 *An American geneticist*: Muller (1932).
144 *Slower than expected*: see Maynard Smith (1989), Chapter 12.
144 *Eliminate bad genes*: Muller (1958).
144 *A good broom*: Trivers (1985), p. 317.
145 *The essence of sex*: Muller (1932).
145 *Without invoking recombination*: Kirkpatrick and Jenkins (1990).

145 *Is not unthinkable*: see Maynard Smith (1988).
146 *Twigs of the phylogenetic descendency tree*: Crow in Michod and Levin (1988), p. 59.
146 *Still need male sperm*: see, for example, Maynard Smith (1988).
146 *Sex can pay its own way*: Maynard Smith (1978).
146 *Immediate advantage*: Kondrashov (1988). See also Nee and Maynard Smith (1990).
147 *Striking metaphor*: Williams (1975).
147 *In 100 different raffles*: see Maynard Smith (1988).
148 *Tangled bank*: Bell (1982).
148 *The phrase is from Darwin*: Darwin (1859), p. 489: 'It is interesting to contemplate an entangled bank, clothed with many plants of many kinds, with birds singing on the bushes, with various insects flitting about, and with worms crawling through the damp earth, and to reflect that these elaborately constructed forms, so different from each other, and dependent on each other in so complex a manner, have all been produced by laws acting around us.'
148 *The fact that seasonal switches*: see, for example, Trivers (1985), p.325.
148 *Red Queen*: van Valen (1973).
148 *Arms races*: Vermeij (1987).
149 *Role of microbial diseases*: Haldane (1949).
149 *Negatively correlated*: Maynard Smith (1978).
149 *A coevolution freak*: Hamilton (1991).
150 *Penny matching*: Hamilton (1980).
150 *Wondrous computer simulations*: Segers and Hamilton (1988) and Hamilton *et al.* (1990). For an early mathematical analysis, see Hutson and Law (1981).
152 *Female mice wrinkle their nose*: Potts *et al.* (1991).
152 *Tierra*: Ray (1991).
152 *Ball floating on water*: Hamilton (1980).
153 *Necessary for health*: see Hamilton (1991).

Chapter 7: notes and references

Richard Dawkins' book *The selfish gene* (1989) is the best introduction to evolutionary game theory. Several chapters in his book *The extended phenotype* (1982), as well as in Robert Trivers' *Social evolution* (1985) and in *An introduction to behavioural ecology* by J. R. Krebs and N. B. Davies (1987) are also devoted to the subject. The book on *Evolution and the theory of games* by John Maynard Smith (1982) is obviously a must for anyone interested in the topic. A sprightly introduction to the theory of games is found in Thomas (1986).

155 *Clever Hans*: see, for example, Boakes (1984).
155 *Fish are no better than birds*: Davis and Perousse (1988).
155 *Six stickleback are jointly served their food*: Milinski (1979).
156 *Capture rate of a hungry fish*: Heller and Milinski (1979).
157 *Elegant trick*: Milinski and Heller (1978).
157 *Battle of the sexes*: Magurran and Nowak (1991).
158 *Killed less frequently*: work by B. H. Seghers (1974, *Evolution*, **28**, 486) quoted in Magurran and Nowak.
159 *Parental investment*: Dawkins (1989), Chapter 9.

159 *Theory of games and economic behaviour*: von Neumann and Morgenstern (1944). The second edition (1947) is still the main reference in the field. For a good introduction, see Luce and Raiffa (1957).
160 *One model of poker*: see, for example, the textbook by Thomas (1986).
160 *Straight into biology*: this approach is from Riechert and Hammerstein (1983).
160 *From lighthearted gambolling to mortal combat*: Huizinga (1970).
161 *Each player has only two strategies*: for an overview, see Rapoport *et al.* (1976).
163 *A classic result*: von Neumann (1928).
163 *US armed forces*: Haywood (1954).
163 *A simplified model of 'Chicken'*: see, for example, Riechert and Hammerstein (1983).
165 *Conflict between deer stags*: see, for example, Krebs and Davies (1987).
165 *This restraint has been much commented upon*: for instance, Lorenz (1966).
166 *To explain it in Darwinian terms*: Maynard Smith and Price (1973).
166 *If most members adopt it*: Maynard Smith (1982).
167 *Pecked to death*: see Lorenz (1982).
167 *Many more aspects*: for a survey, see Maynard Smith (1982).
167 *Playing certain games*: Plato, Laws, vii 803.
167 *Game theory without rationality*: Rapoport (1984).
168 *Playing the field*: see Maynard Smith (1982), p. 23.
168 *French card game*: Fisher (1934).
168 *Playing against nature*: Lewontin (1960).
168 *Unbeatable strategy*: Hamilton (1967).
168 *Game dynamics*: see Taylor and Jonker (1978) and Zeeman (1981). A little-quoted forerunner is Stewart (1971). For a survey, see Hofbauer and Sigmund (1988).
169 *Regret at the static nature*: see von Neumann and Morgenstern (1944), p. 44.
169 *Fictitious play*: Brown (1951).
170 *The same range of motions*: Hofbauer and Sigmund (1988).
170 *Vigilance game*: Pulliam *et al.* (1982).
170 *Rare allele effect*: Haldane (1949).
171 *Rats in a maze*: see Trivers (1985), p. 106.
172 *Fig wasp populations*: Hamilton (1979).
173 *Assess asymmetries*: Maynard Smith and Parker (1976).
174 *No mixed evolutionarily stable strategies*: Selten (1983).
174 *Bluff is transitory*: see Maynard Smith (1982), p. 151.
174 *Bourgeois strategy*: see Maynard Smith and Parker (1976).
174 *Desperado effect*: Grafen (1987).
174 *Speckled wood butterflies*: Davies (1978).
175 *Baboons fighting*: Kummer *et al.* (1964).
175 *Endowment effect*: Thaler (1980).
175 *A spider species*: Burgess (1976).
175 *I would hate to commit myself*: Maynard Smith (1984).
176 *Female ladybirds*: Majerus *et al.* (1983).
176 *A bee's program*: Rothenbuhler (1963).
176 *Field cricket*: Cade (1979).
176 *Human nature*: Wilson (1975); Dawkins (1989, 2nd edn, 1st edn 1976) and Dawkins (1982).
176 *Against any adaptionist explanation*: see Maynard Smith (1982), introduction.

177 *Population geneticists*: see Maynard Smith (1981) and Karlin and Lessard (1986).
177 *Male cowbirds*: King and West (1983).
177 *A duckling learns*: Immelmann (1975).
178 *Rats avoiding food*: Garcia *et al.* (1968).
178 *By a dynamic procedure*: see also Krebs *et al.* (1978).
178 *A candidate mechanism*: Harley (1981). See also Maynard Smith (1984) and the subsequent comments by R. Selten, P. Hammerstein, J. E. Mazur, J. R. Krebs and A. Kacelnik.
178 *Moths fly into candles*: see Maynard Smith (1982), p. 67.
179 *Gaming*: Shubik (1975).
179 *Business games as part of the curricula*: Thomas (1986), p. 224.

Chapter 8: notes and references

The notion of reciprocal altruism made its debut with a seminal paper by Robert Trivers (1971). The little red book on the Iterated Prisoner's Dilemma is, of course, Robert Axelrod's *The evolution of cooperation*. Written in 1984, it is now available from Penguin, with a foreword by Richard Dawkins. Dawkins also included a chapter on this subject, called *Nice guys finish first*, in the new edition of his book *The selfish gene* (1989). Robert Trivers has also a chapter on the evolution of cooperation in his book on *Social evolution* (1985). See also Chapter 9 in Krebs and Davies (1987). For a survey of recent developments, see Axelrod and Dion (1988).

181 *Thousands of scientific publications*: Smale (1980).
181 *How will the rational player act*: see, for example, Rapoport and Chammah (1965) or Mesterton-Gibbons (1992).
182 *In experimental tests*: see Guyer and Perkel (1972) and, for an entertaining account, Chapters 29 and 30 in Hofstadter (1985).
183 *The original tale*: this game was first studied by M. Flood and M. Dresher. For a superb account of its history, see Poundstone (1992).
184 *Shorten the problem step by step*: see Luce and Raiffa (1957), p. 98.
184 *No strategy is always the best reply*: see Axelrod (1984), p. 15.
185 *Inviting them down to a tournament*: see Axelrod (1984), Chapter 2.
185 *Rapoport had pondered*: see Rapoport and Chammah (1965) and Rapoport (1967).
188 *Another tournament*: Axelrod (1980).
190 *There is no looking back*: see Axelrod (1984), p. 64.
190 *Vervet monkeys*: Seyfarth and Cheney (1984).
192 *Accompaniment of random noise*: see Axelrod and Dion (1988) for a survey.
193 *Optimal measure of forgiveness*: Molander (1985).
194 *The cooperation-rewarding zone*: Nowak and Sigmund (1990).
194 *The length of a horn*: Eshel and Motro (1981).
194 *The readiness to cooperate*: Nowak (1990).
195 *If several strategies invade simultaneously*: Boyd and Lorberbaum (1987).
196 *Oscillating compositions*: Nowak and Sigmund (1980).
196 *Representative sample*: May (1987).
196 *Start out with 100 such strategies*: Nowak and Sigmund (1992).
198 *Genetic algorithms*: Axelrod (1987).

200 *With gene duplications*: Lindgren (1991).
201 *Path-breaking paper*: Axelrod and Hamilton (1981). Most of this is reprinted in Chapter 5 of Axelrod (1984).
201 *Stickleback keep abreast*: Milinski (1987).
202 *A strategy based on reciprocity*: Dugatkin (1988).
202 *Among chimpanzees*: see Trivers (1985), p. 362.
202 Au pair *system*: see Krebs and Davies (1987), p. 179.
202 *Subsequent repayment of the debt*: Fisher (1958), pp. 26–7.
202 *Exchange their roles*: Packer (1977).
203 *Behaviour-dependent context*: Feldman and Thomas (1987).
203 *Preferential assortment*: Eshel and Cavalli-Sforza (1982).
203 *Bacteria living in the gut*: see Axelrod (1984), p. 100.
204 *After the cleaner has left*: see Trivers (1985), p. 48.
204 *Signs of injury, illness, or senescence*: see Axelrod and Hamilton (1981),p. 103.
205 *Tragedy of the commons*: Hardin (1968).
205 *And so ground themselves*: Ewald (1987).
206 *A new opportunity to defect*: Axelrod (1986).

Chapter 9: notes and references

The evolutionary relevance of animal play is well worked out in Fagen (1981). The essay of Huizinga (1970) on *Homo Ludens* shows how much of human culture is rooted in play.

207 *Neotenic properties*: see, for example, Gould (1982), Chapter 7: The child as man's real father.
207 *An instance of self-domestication*: see, for example, Lorenz (1965) and Chapter 6 of Lorenz (1978).
207 *Widespread among higher animals*: Fagen (1981).
208 *Playing behaviour of seals*: see Hassenstein (1980), p. 115.
208 *The most playful animal species*: Fagen (1981).
209 *Philosophical investigations*: Wittgenstein (1953).
210 *Homo Ludens*: Huizinga (1970).
210 *Stock exchange and random walks*: the idea was first explored in the 1900 thesis of Louis Bachelier. For a general overview, see Malkiel (1985). A good introduction is Chapter 4 in Casti (1990).
210 *Jack Dempsey*: see Chapter 7 of Lorenz (1966).
211 *Fighting and Wooing*: Lorenz (1982).
211 *Assembled under the light*: Le Corbusier (1927).
212 *Quantum reality*: for an overview of the debate on quantum reality, see Casti (1989), Chapter 7.
212 *Philosopher's stone*: Boltzmann (reprinted 1979), p. 27.
212 *Mathematician's newspeak*: for a discussion, see, for example, Davis and Hersh (1981).
213 *Form a square*: see Stewart (1991).
213 *Helping to make war*: Hardy (1940).
213 *Public-key methods*: see, for example, Stewart (1987), Chapter 2.
214 *Motivated rather than applied*: Aubin (1991).
214 *Not seriousness*: Freud (1908).

Bibliography

Alexander, R. D., Noonan, K. M., and Crespi, B. J. (1991). The evolution of eusociality. In *The biology of the naked mole rat* (ed. P. W. Sherman, J. U. M. Jarvis, and R. D. Alexander), pp. 3–44. Princeton University Press.

Anderson, R. M. and May, R. M. (1991). *Infectious diseases of humans: dynamics and control*. Oxford University Press.

Arthur, B. (1990). Positive feedbacks in economy. *Scientific American*, **262**(2), 80–5.

Aubin, J. P. (1991). *Viability theory*. Birkhäuser, Basel.

Axelrod, R. (1980). Effective choice in the prisoner's dilemma. *Journal of Conflict Resolution*, **24**, 3–25.

Axelrod, R. (1984). *The evolution of cooperation*. Basic Books, New York (also published 1991, Penguin, Harmondsworth).

Axelrod, R. (1986). An evolutionary approach to norms. *American Political Science Review*, **80**, 1095–111.

Axelrod, R. (1987). The evolution of strategies in the iterated prisoner's dilemma. In *Genetic algorithms and simulated annealing* (ed. D. Davis), pp. 32–43, Pitman, London.

Axelrod, R. and Dion, D. (1988). The further evolution of cooperation. *Science*, **242**, 1385–90.

Axelrod, R. and Hamilton, W. D. (1981). The evolution of cooperation. *Science*, **211**, 1390–6.

Balmford, A. and Read, A. F. (1991). Testing alternative models of sexual selection through female choice. *Trends in Ecology and Evolution*, **6**, 274–6.

Barton, N. H. and Turelli, M. (1989). Evolutionary quantitative genetics: how little do we know. *Annual Review of Genetics*, **23**, 337–70.

Bartz, S. (1979). Evolution of eusociality in termites. *Proceedings of the National Academy of Sciences USA*, **76**, 5764–8.

Bell, G. (1982). *The masterpiece of nature: the evolution and genetics of sexuality*. University of California Press, Berkeley.

Bennett, J. H. (1983). *Natural selection, heredity, and eugenics*. Clarendon Press, Oxford.

Bergerud, A. T. (1983). Prey switching in simple ecosystems. *Scientific American*, **249**(6), 116–24.

Berlekamp, E., Conway, J., and Guy, R. (1982). *Winning ways for your mathematical plays*. Academic Press, New York.

Boakes, R. (1984). *From Darwin to behaviourism: psychology and the mind of animals*. Cambridge University Press.

Boltzmann, L. (1905) (1979). *Populäre Schriften* (ed. E. Broda). Vieweg, Braunschweig.

Box, J. (1978). *R. A. Fisher: the life of a scientist*. Wiley, New York.

Boyd, R. and Lorberbaum, J. P. (1987). No pure strategy is evolutionarily stable in the repeated Prisoner's Dilemma. *Nature*, **327**, 58–9.

Brown, G. W. (1951). Iterative solutions of games by fictitious play. In *Activity analysis of production and allocation* (ed. T. C. Koopmans), pp. 374–6. Wiley, New York.

Burgess, J. W. (1976). Social spiders. *Scientific American*, **234**(3), 100–6.

Burks, A. (ed.) (1970). *Essays on cellular automata*. University of Illinois Press, Urbana.

Buss, L. W. (1987). *The evolution of individuality*. Princeton University Press.

Byl, J. (1989). Self-reproduction in small cellular automata. *Physica D*, **35**, 295–9.

Cade, W. (1981). Alternative male strategies: genetic differences in crickets. *Science*, **212**, 563–4.

Cairns-Smith, J. (1982). *Seven clues for the origin of life*. Cambridge University Press.

Cann, R. L., Stoneking, M., and Wilson, A. C. (1987). Mitochondrial DNA and human evolution. *Nature*, **325**, 31–6.

Casti, J. (1989). *Paradigms lost*. Morrow, New York.

Casti, J. (1990). *Searching for certainty*. Morrow, New York.

Cavalli-Sforza, L. L. and Bodmer, W. F. (1971). *The genetics of human populations*. Freeman, New York.

Charlesworth, B. (1978). The population genetics of anisogamy. *Journal of Theoretical Biology*, **73**, 347–57.

Charnov, E. L. (1982). *The theory of sex allocation*. Princeton University Press.

Clark, R. W. (1969). *JBS: the life and work of J. B. S. Haldane*. Coward-McCann, New York.

Codd, E. F. (1968). *Cellular automata*. Academic Press, New York.

Cohen, J. (1976). Irreproducible results and the breeding of pigs. *Biosciences*, **26**, 391–4.

Cohen, J. (1989). Food webs and community structure. In *Perspectives in ecology* (ed. J. Roughgarden, R. M. May and S. A. Levin), pp. 181–202. Princeton University Press.

Colinvaux, P. (1978). *Why big fierce animals are rare*. Princeton University Press.

Conover, D. O. and Voorhees, D. A. (1990). Evolution of a balanced sex-ratio by frequency-dependent selection in a fish. *Science*, **250**, 1556–8.

Le Corbusier (1927). *Towards a new architecture*. Architectural Press, London. Reprinted 1986 by Dover, New York.

Crow, J. F. (1979). Genes that violate Mendel's rules. *Scientific American*, **240**(2), 104–13.

Crow, J. F. (1986). *Basic concepts in population, quantitative and evolutionary genetics*. Freeman, New York.

Crow, J. F. and Kimura, M. (1970). *An introduction to population genetics*. Harper and Row, New York.

Darwin, C. (1859). *The origin of species* (reprinted 1964). Harvard University Press, Cambridge, MA.

Darwin, F. (ed.) (1887). *The life and letters of Charles Darwin*, vol. I. Murray, London.

Davies, N. B. (1978). Territorial defence in the speckled wood butterfly. *Animal Behaviour*, **26**, 138–47.

Davis, P. J. and Hersh, R. (1981). *The mathematical experience*. Birkhäuser, Boston.

Davis, H. and Perousse, R. (1988). Numerical competence in animals. *Behavioural and Brain Sciences*, **11**, 561–615.

Dawkins, R. (1982). *The extended phenotype*. Freeman, Oxford.

Dawkins, R. (1986). *The blind watchmaker*. Longman, Harlow.
Dawkins, R. (1976, 1989). *The selfish gene* (2nd edn). Oxford University Press.
DeBeer, G. (1964). *Charles Darwin: a scientific biography*. Doubleday, New York.
Devaney, R. L. (1986). *An introduction to chaotic dynamical systems*. Benjamin-Cummings, Menlo Park.
Dewdney, A. K. (1984). Sharks and fish wage war on the planet Wa-Tor. *Scientific American*, **251**(6), 14–22.
Dewdney, A. K. (1989). *The Turing omnibus*. Computer Science Press, New York.
Dowling, W. (1990). Computer viruses: diagonalization and fixed points. *Notices of the American Mathematical Society*, **37**, 858–60.
Dugatkin, L. A. (1988). Do guppies play Tit For Tat during predator inspection visits? *Behavioural Ecology Sociobiology*, **23**, 395–9.
Dyson, F. (1988). *Infinite in all directions*. Harper and Row, New York.
Edey, M. A. and Johanson, D. C. (1990). *Blueprints: Solving the mystery of evolution*. Oxford University Press.
Eigen, M. (1971). Self-organization of matter and the evolution of biological macro-molecules. *Naturwissenschaften*, **58**, 465–526.
Eigen, M. and Schuster, P. (1979). *The hypercycle. A principle of natural self-organization*. Springer, Berlin.
Eigen, M., Gardiner, W., Schuster, P., and Winkler-Oswatitsch, R. (1981). The origin of genetic information. *Scientific American*, **244**(4), 78–95.
Ekeland, I. (1988). *Mathematics and the unexpected*. University of Chicago Press.
Ekeland, I. (1991). *Au hasard*. Edition du Seuil, Paris (in French).
Eshel, I. and Cavalli-Sforza, L. (1982). Assortment of encounters and evolution of cooperativeness. *Proceedings of the National Academy of Sciences USA*, **79**, 1331–5.
Eshel, I. and Motro, A. (1981). Kin selection and strong evolutionary stability of mutual help. *Theoretical Population Biology*, **19**, 420–33.
Ewald, P. W. (1987). Transmission modes and evolution of the parasitism–mutualism continuum. *Annals of the New York Academy of Sciences*, **503** (Endocytobiology III), 295–306.
Fagen, R. M. (1981). *Animal play behaviour*. Oxford University Press.
Falconer, D. S. (1981). *Introduction to quantitative genetics* (2nd edn). Longman, Harlow.
Feldman, M. and Thomas E. (1987). Behaviour-dependent contexts for repeated plays of the Prisoner's Dilemma II. *Journal of Theoretical Biology*, **128**, 297–315.
Feller, W. (1968). *An introduction to probability theory* (3rd edn). Wiley, New York.
Fisher, R. A. (1918). The correlation between relatives on the supposition of Mendelian inheritance. *Transactions of the Royal Society of Edinburgh*, **52**, 379–433.
Fisher, R. A. (1934). Randomisation and an old enigma of card play. *Mathematical Gazette*, **18**, 294–7.
Fisher, R. A. (1936). Has Mendel's work been re-discovered? *Annals of Science*, **i**, 115–37.
Fisher, R. A. (1958). *The genetical theory of natural selection* (2nd edn). Dover, New York.
Freud, S. (1908). *Poetry and phantasizing*. (Collected Papers, Vol. 7.) The International Psycho-Analytical Library, Hogarth Press, London.

Garcia, J., McGowan, B. K., Ervin, F. R., and Koelling, R. A. (1968). Cues: their relative effectiveness as a function of the reinforcer. *Science*, **196**, 794–6.

Gardner, M. (1970). Mathematical games: the fantastic combinations of John Conway's new Solitary game 'Life'. *Scientific American*, **223**, 120–3.

Gardner, M. (1983). *Wheels, life, and other amusements*. Freeman, New York.

Gillespie, J. H. (1991). *The causes of molecular evolution*, Oxford University Press, New York.

Gilpin, M. E. (1973). Do hares eat lynx? *American Naturalist*, **107**, 727–30.

Gilpin, M. E. (1979). Spiral chaos in a predator-prey model. *American Naturalist*, **113**, 306–8.

Gleick, J. (1987). *Chaos: the making of a new science*. Viking Press, New York.

Gould, S. J. (1979). Shades of Lamarck. *Natural History*, **88**, 212–28.

Gould, S. J. (1982). *Ever since Darwin*. Penguin, Harmondsworth.

Gould, S. J. (1983). *Hen's teeth and horse's toes*. Penguin, Harmondsworth.

Grafen, A. (1987). The role of divisively asymmetric contests: respect for ownership and the desperado effect. *Animal Behaviour*, **35**, 462–7.

Gruber, H. E. and Barrett, P. H. (1974). *Darwin on man: a psychological study of scientific creativity, together with Darwin's early and unpublished notebooks*. Dutton, New York.

Guyer, M. and Perkel, B. (1972). *Experimental games: A bibliography of 1945–1971*. Communication No. 293, Mental Health Research Institute, University of Chicago.

Haig, D. and Grafen, A. (1991). Genetic scrambling as a defence against meiotic drive. *Journal of Theoretical Biology*, **953**, 531–58.

Haldane, J. B. S. (1927). *Possible worlds*. Chatto and Windus, London.

Haldane, J. B. S. (1932). *The causes of evolution* (reprinted 1966). Princeton University Press.

Haldane, J. B. S. (1949). Disease and evolution. *Supplemente a la Ricerca Scientifica*, Anno **19**, 68–75.

Haldane, J. B. S. (1964). In defense of beanbag genetics. *Perspectives in Biology and Medicine*, **7**, 343–59.

Hall, N. (ed.) (1992). *The* New Scientist *guide to chaos*. Penguin, Harmondsworth.

Hamilton, W. D. (1964). The genetical evolution of social behaviour. *Journal of Theoretical Biology*, **7**, 1–52.

Hamilton, W. D. (1967). Extraordinary sex ratios. *Science*, **156**, 477–88.

Hamilton, W. D. (1978). Evolution and diversity under bark. *Symposium of the Royal Society of Entomology, London*, **9**, 154–75.

Hamilton, W. D. (1979). Wingless and fighting males in fig wasps and other insects. In *Sexual selection and reproductive competition* (ed. M. S. and N. A. Blum). Academic Press, London.

Hamilton, W. D. (1980). Sex versus non-sex versus parasites. *Oikos*, **35**, 282–90.

Hamilton, W. D. (1991). Memes of Haldane and Jayakar in a theory of sex. *Journal of Genetics*, **69**, 17–32.

Hamilton, W. D. and Zuk, M. (1982). Heritable true fitness and bright birds: a role for parasites? *Science*, **218**, 384–7.

Hamilton, W. D., Axelrod, R., and Tanese, R. (1990). Sexual reproduction as an adaptation to resist parasites. *Proceedings of the National Academy of Sciences USA*, **87**, 3566–73.

Hardin, G. (1968). The tragedy of the commons. *Science*, **162**, 1243–8.

Hardy, G. H. (1908). Mendelian proportions in a mixed population. *Science*, **28**, 49–50.

Hardy, G. H. (1940). *A mathematician's apology*. Cambridge University Press.

Harley, B. (1981). Learning the evolutionarily stable strategy. *Journal of Theoretical Biology*, **89**, 611–33.

Hartl, D. L. (1980). *Principles of population genetics*. Sinauer, Sunderland.

Hassell, M. P., Comins, H., and May, R. M. (1991). Spatial structure and chaos in insect population dynamics. *Nature*, **353**, 255–8.

Hassenstein, B. (1980). *Instinkt, Lernen, Spielen, Einsicht*. Piper und Co., München (in German).

Hastings, A. and Powell, T. (1991). Chaos in a three-species food chain. *Ecology*, **72**, 896–903.

Haywood, O. G. (1954). Military decisions and game theory. *Journal of Operations Research Society of America*, **2**, 365–85.

Heims, S. J. (1980). *John von Neumann and Norbert Wiener*. MIT Press, Cambridge, MA.

Heller, R. and Milinski, M. (1979). Optimal foraging of sticklebacks on swarming prey. *Animal Behaviour*, **27**, 1127–41.

Hodges, A. (1983). *Alan Turing: the enigma*. Simon and Schuster, New York.

Hoekstra, R. (1987). The evolution of sexes. In *The evolution of sex and its consequences* (ed. S. C. Stearns), pp. 59–92. Birkhäuser, Basel.

Hoekstra, R., Iwasa, Y., and Weissing, F. J. (1991). The origin of isogamous sexual differentiation, in Selten, R. (ed.), *Game equilibrium models I*, pp. 155–81, Springer, Berlin.

Hofbauer, J. and Sigmund, K. (1988). *The theory of evolution and dynamical systems*. Cambridge University Press.

Hofstadter, D. R. (1980). *Gödel, Escher, Bach: An eternal golden braid*. Penguin, Harmondsworth.

Hofstadter, D. R. (1985). *Metamagical themas*. Penguin, Harmondsworth.

Huffaker, C. B. (1958). Experimental studies on predation: dispersion factors and predator–prey interactions. *Hilgardia*, **27**, 343–83.

Huizinga, J. (1970). *Homo Ludens*. Paladin, London.

Hurst, L. (1992). Intra-genomic conflict as an evolutionary force. *Proceedings of the Royal Society (London)*, Series B **248**, 135–41.

Hurst, L. and Hamilton, W. D. (1992). Cytoplasmic fusion and the nature of the sexes. *Proceedings of the Royal Society (London)*, Series B **247**, 189–95.

Hutchinson, G. E. (1965). *The ecological theatre and the evolutionary play*. Yale University Press, New Haven.

Hutson, V. (1984). Predator-mediated coexistence with a switching predator. *Mathematical Biosciences*, **68**, 233–46.

Hutson, V. and Vickers, G. T. (1983). A criterion for permanent coexistence of species with an application to a two-prey one-predator system. *Mathematical Biosciences*, **63**, 253–69.

Iltis, H. (1932). *Life of Mendel*. Norton, New York.

Immelmann, K. (1975). Ecological significance of imprinting and early learning. *Annual Review of Ecology and Systematics*, **6**, 15–37.

Jacob, F. (1976). *The logic of life*. Random, New York.

Jacob, F. (1982). *The possible and the actual*. University of Washington Press, Seattle.

Jacquard, A. (1974). *The genetic structure of populations*. Springer, New York.

Jacquard, A. (1990). *Inventer l'homme*. Edition Complexe, Paris (in French).

Jagers, P. (1975). *Branching processes with biological applications*. Wiley, London.

Jansen, V. and Sabelis, M. V. (1992). Prey dispersal and predator persistence. *Experimental and Applied Acarology*, **14**, 215–31.

Karlin, S. (1966). *A first course in stochastic processes*. Academic Press, New York.

Karlin, S. and Feldman, M. (1970). Linkage and selection: two locus symmetric viability model. *Theoretical Population Biology*, **1**, 37–71.

Karlin, S. and Lessard, S. (1986). *Sex ratio evolution*. Princeton University Press.

Kimura, M. (1979). The neutral theory of molecular evolution. *Scientific American*, **241**(5), 94–104.

Kimura, M. (1983). *The neutral theory of molecular evolution*. Cambridge University Press.

Kimura, M. and Ohta, T. (1971). *Theoretical aspects of population genetics*. Princeton University Press.

King, A. P. and West, M. J. (1983). Epigenesis of cowbird songs—a joint endeavour of males and females. *Nature*, **305**, 704–6.

Kingsland, S. (1985). *Modeling nature; episodes in the history of population ecology*. University of Chicago Press.

Kirkpatrick, M. (1982). Sexual selection and the evolution of female choice. *Evolution*, **36**, 1–12.

Kirkpatrick, M. and Jenkins, C. D. (1990). Genetic segregation and the maintenance of sexual reproduction. *Nature*, **339**, 300–1.

Kirlinger, G. (1986). Permanence in Lotka–Volterra equations: linked predator-prey systems. *Mathematical Biosciences*, **82**, 165–91.

Kondrashov, A. S. (1988). Deleterious mutations and the evolution of sexual reproduction. *Nature*, **336**, 435–40.

Koth, M. and Kemler, F. (1986). A one-locus two-allele selection model admitting stable limit cycles. *Journal of Theoretical Biology*, **122**, 263–7.

Krebs, J. R. and Davies, N. B. (1987). *An introduction to behavioural ecology* (2nd edn). Blackwell, Oxford.

Krebs, J. R., Kacelnik, A., and Taylor, P. (1978). Test of optimal sampling by foraging great tits. *Nature*, **275**, 27–31.

Kummer, H., Götz, W., and Angst, W. (1964). Triadic differentiation: an inhibitory process protecting pair bonds in baboons. *Behaviour*, **41**, 1–26.

Laing, R. (1977). Automaton models of reproduction by self-inspection. *Journal of Theoretical Biology*, **66**, 437–56.

Lande, R. (1981). Models of speciation by sexual selection on polygenic traits. *Proceedings of the National Academy of Sciences USA*, **78**, 3721–5.

Langton, C. G. (1984). Self-reproduction in cellular automata. *Physica D*, **10**, 134–43.

Langton, C. G. (1986). Studying artificial life with cellular automata. *Physica D*, **22**, 120–49.

Langton, C. G. (1989). *Artificial life I*, Proceedings of the first workshop on the synthesis and simulation of living systems, Vol. VI, Santa Fe Institute for Studies in the Science of Complexity.

Langton, C. G., Taylor, C., Farmer, J. D., and Rasmussen, S. (ed.) (1991). *Artificial life II*, Proceedings of the second workshop on the synthesis and simulation of living systems, Vol. X, Santa Fe Institute for Studies in the Science of Complexity.

Legendi, T. and Szentivanyi, T. (ed.) (1983). *Leben und Werk von John von Neumann*. Bibl. Inst. Mannheim.

Lewontin, R. C. (1960). Evolution and the theory of games. *Journal of Theoretical Biology*, **1**, 382–403.

Lindgren, K. (1991). Evolutionary phenomena in simple dynamics. In *Artificial life II* (ed. C. G. Langton, C. Taylor, J. D. Farmer, and S. Rasmussen), Proceedings of the second workshop on the synthesis and simulation of living systems, Vol. X, pp. 295–312, Santa Fe Institute for Studies in the Science of Complexity.

Lorenz, K. (1965). *Evolution and modification of behaviour*. Chicago University Press.

Lorenz, K. (1966). *On Aggression*. Harcourt-Brace-Jovanovich, New York.

Lorenz, K. (1978). *Civilized Man's eight deadly sins*. Harcourt-Brace-Jovanovich, New York.

Lorenz, K. (1982). *King Solomon's ring*. Harper-Collins, New York.

Lotka, A. J. (1923). Contributions to the analysis of malaria epidemiology. *American Journal of Hygiene*, **3**, 1–121 (in 5 parts; Part 4 jointly with F. R. Sharpe).

Lotka, A. J. (1931). The extinction of families II. *Journal of the Washington Academy of Sciences*, **21**, 453–9.

Luce, R. D. and Raiffa, H. (1957). *Games and decisions*. Wiley, New York.

MacArthur, R. H. (1958). Population ecology of some warblers of northeastern coniferous forests. *Ecology*, **39**, 599–619.

MacArthur, R. H. (1966). The theory of niche. In *Population Biology and Evolution* (ed. R. Lewontin), pp. 152–66. Syracuse University Press.

MacArthur, R. H. (1972). Coexistence of species. In *Challenging biological problems* (ed. J. Behnke), pp. 257–72. Oxford University Press.

Magurran, A. E. and Nowak, M. (1991). Another battle of the sexes. *Proceedings of the Royal Society London*, **246**, 31–8.

Majerus, M., O'Donald, P., and Weir, J. (1983). Female mating preference is genetic. *Nature*, **300**, 521–3.

Malkiel, B. (1985). *A random walk down Wall Street* (4th edn). Norton, New York.

May, R. M. (1976). Simple mathematical models with very complicated dynamics. *Nature*, **261**, 459–67.

May, R. M. (1978). The evolution of ecological systems. *Scientific American*, **239**(3), 119–33.

May, R. M. (1987). More evolution of cooperation. *Nature*, **327**, 15–17.

May, R. M. and Leonard, W. (1975). Nonlinear aspects of competition. *SIAM Journal of Applied Mathematics*, **29**, 243–52.

Maynard Smith, J. (1971). The origin and maintenance of sex. In *Group selection* (ed. G. C. Williams), pp. 163–75. Aldine-Atherton, Chicago.

Maynard Smith, J. (1974). The theory of games and the evolution of animal conflicts. *Journal of Theoretical Biology*, **47**, 209–21.

Maynard Smith, J. (1978). *The evolution of sex*. Cambridge University Press.

Maynard Smith, J. (1981). Will a sexual population evolve to an ESS? *American Naturalist*, **177**, 1015–18.

Maynard Smith, J. (1982). *Evolution and the theory of games*. Cambridge University Press.

Maynard Smith, J. (1984). Game theory and the evolution of behaviour. *Behavioural and Brain Sciences*, **7**, 95–125.

Maynard Smith, J. (1988). *Games, sex and evolution*. Harvester Wheatsheaf, New York.

Maynard Smith, J. (1989). *Evolutionary genetics*. Oxford University Press.

Maynard Smith, J. and Parker, G. A. (1976). The logic of asymmetric contests. *Animal Behaviour*, **24**, 159–75.

Maynard Smith, J. and Price, G. R. (1973). The logic of animal conflict. *Nature*, **246**, 15–18.

Mayr, E. (1963). *Animal species and evolution*. Harvard University Press.

Mendel, G. (1865). Versuche über Pflanzen-Hybriden. *Verhandlungen des natur-forschenden Vereins in Brünn*, **10**, 3–47.

Mesterton-Gibbons, M. (1992). *An introduction to game-theoretic modelling*. Addison Wesley, Redwood City.

Michod, R. E. and Levin, B. R. (ed.) (1988). *The evolution of sex: an examination of current ideas*. Sinauer, Sunderland.

Milinski, M. (1979). An evolutionarily stable feeding strategy in stickleback. *Zeitschrift für Tierpsychologie*, **51**, 36–40.

Milinski, M. (1987). Tit For Tat in stickleback and the evolution of cooperation. *Nature*, **325**, 434–5.

Milinski, M. and Bakker, T. C. (1990). Female stickleback use male coloration in mate choice and hence avoid parasitized males. *Nature*, **344**, 330–3.

Milinski, M. and Heller, R. (1978). Influence of a predator on the optimal foraging behaviour of stickleback. *Nature*, **275**, 642–4.

Molander, P. (1985). The optimal level of generosity in a selfish, uncertain environment. *Journal of Conflict Resolution*, **29**, 611–15.

Monod, J. (1972). *Chance and necessity*. Fontana, London.

Moran, P. A. P. (1953). The statistical analysis of the Canadian lynx cycle I. *Australian Journal of Zoology*, **1**, 163–73.

Muller, H. J. (1932). Some genetic aspects of sex. *American Naturalist*, **66**, 118–38.

Muller, H. J. (1958). Evolution by mutation. *Bulletin of the American Mathematical Society*, **64**, 137–60.

Muratori, S. and Rinaldi, S. (1992). Low and high frequency oscillations in three-dimensional food chains. *SIAM Journal in Applied Mathematics* (in press).

Murray, J. D. (1989). *Mathematical biology*. Springer, Berlin–Heidelberg–New York.

Müller-Hill, B. (1988). *Murderous science*. Oxford University Press.

Nee, S. and Maynard Smith, J. (1990). The evolutionary biology of molecular parasites. *Parasitology*, **100**, S5–S18.

von Neumann, J. (1928). Zur Theorie der Gesellschaftsspiele. *Mathematische Annalen*, **100**, 295–320.

von Neumann, J. (1966). *Theory of self-reproducing automata*. University of Illinois Press, Urbana.

von Neumann, J. and Morgenstern, I. (1944). *Theory of games and economic behaviour*. Princeton University Press.

Nicholson, A. J. (1954). An outline of the dynamics of animal populations. *Australian Journal of Zoology*, **2**, 9–65.

Nowak, M. (1990). An evolutionarily stable strategy may be inaccessible. *Theoretical Population Biology*, **142**, 237–41.

Nowak, M. and Sigmund, K. (1990). The evolution of stochastic strategies in the prisoner's dilemma. *Acta Applicandae Mathematicae*, **20**, 247–65.

Nowak, M. and Sigmund, K. (1992). Tit For Tat in heterogeneous populations. *Nature*, **355**, 250–3.

O'Donald, P. (1982). *Genetic models of sexual selection*. Cambridge University Press.

Orel, V. (1984). *Mendel*. Oxford University Press.

Packer, C. (1977). Reciprocal altruism in *Papio anubis*. *Nature*, **265**, 441–3.

Paine, R. T. (1966). Food web complexity and species diversity. *American Naturalist*, **100**, 65–75.

Paine, R. T. (1992). Food web analysis through field measurements of per capita interaction strength. *Nature*, **355**, 73–7.

Parker, G., Baker, R., and Smith, V. (1972). The origin and evolution of gamete dimorphism and the male–female phenomenon. *Journal of Theoretical Biology*, **36**, 529–53.

Partridge, L. (1983). Mate choice increases a component of offspring fitness in fruit flies. *Nature*, **283**, 290–1.

Pomianowski, A., Iwasa, Y., and Nee, S. (1991). The evolution of costly mate preference. *Evolution*, **45**, 1422–30.

Potts, W. K., Manning, C. J., and Wakeland, E. K. (1991). Mating patterns in seminatural populations of mice influenced by MHC genotype. *Nature*, **352**, 619–21.

Poundstone, W. (1987). *The recursive universe*. Oxford University Press.

Poundstone, W. (1992). *The Prisoner's Dilemma*. Doubleday, New York.

Pratt, D. M. (1943). Analysis of population development in *Daphnia* at different temperatures. *Biological Bulletin*, **85**, 116–40.

Proctor, R. M. (1988). *Racial hygiene: medicine under the Nazis*. Harvard University Press, Cambridge, MA.

Provine, W. (1971). *The origins of theoretical population genetics*. University of Chicago Press.

Provine, W. (1986). *Sewall Wright and evolutionary biology*. University of Chicago Press.

Pulliam, H. R., Pyke, G. H., and Caraco, T. (1982). The scanning behaviour of juncos: a game theoretical approach. *Journal of Theoretical Biology*, **95**, 89–103.

Rapoport, A. (1967). Escape from paradox. *Scientific American*, **217**(1), 50–6.

Rapoport, A. (1984). Game theory without rationality. *Behavioural and Brain Sciences*, **7**, 114.

Rapoport, A. and Chammah, A. (1965). *The Prisoner's Dilemma*. University of Michigan Press, Ann Arbor.

Rapoport, A., Guyer, M., and Gordon, D. (1976). *The 2 × 2 game*. University of Michigan Press, Ann Arbor.

Raup, D. M. and Sepkovski, J. J., Jr. (1982). Mass extinction in the marine fossil record. *Science*, **215**, 1501–3.

Ray, T. S. (1991). An approach to the synthesis of life. In *Artificial life II* (ed. C. G. Langton, C. Taylor, J. D. Farmer, and S. Rasmussen). Proceedings of the second workshop on the synthesis and simulation of living systems. Vol. X. Sante Fe Institute for Studies in the Science of Complexity.

Real, L. A. and Brown, J. H. (ed.) (1991). *Foundations of ecology: classic papers with comments*. University of Chicago Press.

Rennie, J. (1992). Living together. *Scientific American*, **266**(1), 104–13.

Renshaw, E. (1991). *Modelling biological populations in space and time*. Cambridge University Press.

Ridley, M. (1985). *The problems of evolution*. Oxford University Press.

Riechert, S. and Hammerstein, P. (1983). Game theory in the ecological context. *Annual Review of Ecology and Systematics*, **14**, 377–409.

Riedl, R. (1979). *The order of living organisms*. Wiley, New York.

Rodgers, J. (1991). Mechanisms Mendel never knew. *Mosaic*, **22**, 2–11.

Rothenbuhler, W. C. (1963). Behaviour genetics of nest cleaning in honey bees IV. *American Zoologist*, **4**, 111–23.

Roughgarden, J. (1979). *Theory of population genetics and evolutionary ecology*. MacMillan, New York.

Roughgarden, J., May, R. M., and Levin, S. A. (ed.) (1989). *Perspectives in ecology*. Princeton University Press.

Schaffer, W. M. (1985). Order and chaos in ecological systems. *Ecology*, **66**, 93–100.

Schuster, P. and Sigmund, K. (1989). Fixation probabilities for advantageous mutants. *Mathematical Biosciences*, **95**, 37–51.

Scudo, F. M. (1970). Vito Volterra and theoretical ecology. *Theoretical Population Biology*, **2**, 1–23.

Scudo, F. and Ziegler, J. (1978). *The golden age of theoretical ecology: 1923–1940*. Lecture notes in biomathematics, Springer.

Segers, J. and Hamilton, W. D. (1988). Parasites and sex. In *The evolution of sex: an examination of current ideas* (ed. R. E. Michod and B. R. Levin), pp. 176–93. Sinauer, Sunderland.

Selten, R. (1983). A note on evolutionarily stable strategies in asymmetrical animal conflicts. *Journal of Theoretical Biology*, **84**, 93–101.

Seyfarth, R. M. and Cheney, D. L. (1984). Grooming, alliances and reciprocal altruism in vervet monkeys. *Nature*, **308**, 541–3.

Shapiro, R. (1988). *Origins*. Penguin, Harmondsworth.

Sheppard, P. M. (1958). *Natural selection and heredity*. Hutchinson, London.

Sherman, P. W., Jarvis, J. U. M., and Alexander, R. D. (ed.) (1991). *The biology of the naked mole rat*. Princeton University Press.

Shubik, M. (1975). *The uses and methods of gaming*. Elsevier, New York.

Simberloff, D. S. and Wilson, E. O. (1969). Experimental zoogeography of islands: the colonization of empty islands. *Ecology*, **50**, 278–96.

Smale, S. (1980). The prisoner's dilemma and dynamical systems associated to non-cooperative games. *Econometrica*, **48**, 1617–34.

Sorensen, R. (1991). Thought experiments. *American Scientist*, **79**, 250–63.

Stearns, S. C. (ed.) (1987). *The evolution of sex and its consequences*. Birkhäuser, Basel.

Stewart, F. M. (1971). Evolution in a predator–prey model. *Theoretical Population Biology*, **1**, 493–506.

Stewart, I. (1987). *The problems of mathematics*. Oxford University Press.

Stewart, I. (1989). *Does God play dice?* Blackwell, Oxford.

Stewart, I. (1991). The ultimate jigsaw puzzle. *New Scientist*, **135**, 30–3.

Sved, J. A and Mayo, O. (1970). The evolution of dominance. In *Mathematical topics in population genetics* (ed. K. Kojima), pp. 289–316. Springer, Berlin.

Szekely, I. (1986). *Paradoxes in probability theory and mathematical statistics*. Reidel, Dordrecht.

Tanner, J. T. (1975). The stability and intrinsic growth rates of prey and predator populations. *Ecology*, **56**, 855–67.

Taylor, P. and Jonker, L. (1978). Evolutionarily stable strategies and game dynamics. *Mathematical Biosciences*, **40**, 145–56.

Templeton, A. (1992). Human origins and analysis of mitochondrial DNA-sequences. *Science*, **255**, 737.

Thaler, R. (1980). Toward a positive theory of consumer choice. *Economics of Behaviour and Organization*, **1**, 39–60.

Thomas, L. C. (1986). *Games, theory and applications*. Ellis Horwood, Chichester.

Trivers, R. (1971). The evolution of reciprocal altruism. *Quarterly Review of Biology*, **46**, 35–57.

Trivers, R. (1985). *Social evolution*. Benjamin Cummings, Menlo Park.

Turing, A. M. (1936–37). On computable numbers, with an application to the *Entscheidungsproblem*. *Proceedings of the London Mathematical Society*, **2**(42), 230–65.

Ulam, S. M. (1970). On some mathematical problems connected with patterns of growth of figures. In *Essays on cellular automata* (ed. A. Burks), pp. 214–28. University of Illinois Press, Urbana.

Ulam, S. M. (1976). *Adventures of a mathematician*. Charles Scribner's Sons, New York.

van Valen, L. (1973). A new evolutionary law. *Evolutionary Theory*, **1**, 1–30.

Vermeij, G. J. (1987). *Evolution and escalation: an ecological history of life*. Princeton University Press.

Wagner, G. (1990). The domestication of replicators. *Evolutionary Trends in Plants*, **4**, 71–2.

Wagner, G. and Bürger, R. (1985). On the evolution of dominance modifiers II. *Journal of Theoretical Biology*, **113**, 475–500.

Weinberg, W. (1908). Über einen Nachweis der Vererbung beim Menschen. *Jahreshefte des Vereins für vaterländische Naturkunde in Württemberg*, **64**, 369–82.

Weismann, A. (1904). *On evolution*. Edward Arnold, London.

Weissing, F. J. (1991). Evolutionary stability and dynamic stability in a class of evolutionary normal form games. In *Game equilibrium models I* (ed. R. Selten), pp. 29–96. Springer, Berlin.

Williams, G. C. (1975). *Sex and evolution*. Princeton University Press.

Wilson, E. O. (1975). *Sociobiology: the new synthesis*. Harvard University Press, Cambridge, MA.

Wittgenstein, L. (1953). *Philosophical investigations*. Blackwell, Oxford.

Wolfram, S. (ed.) (1986). *Theory and applications of cellular automata*. World Scientific, Singapore.

Wright, S. (1968, 1969, 1977, 1978). *Evolution and the genetics of populations*, 4 vols. University of Chicago Press.

Zeeman, E. C. (1981). Dynamics of the evolution of animal conflicts. *Journal of Theoretical Biology*, **89**, 249–70.

Index

adaptation 84, 86–8, 127, 143–4, 147–9, 152, 166, 170, 176
advantage, selective 82–6, 89, 107, 110, 113–14, 125, 128, 146, 165
aggression 166, 176
albinism 106, 121
Alexander, R. D. 117–18
algorithm, genetic 198–200
allele 102, 104, 106–7, 116, 176–7
  deleterious 111, 115
  dominant 97–8, 106–8, 121
  rare allele effect 89, 170
  recessive 106–8, 111–12, 114, 115
allometric limitation 73
amino acid 19, 84, 89–90
ant 119, 204
arms race 148–51
assessment 165, 173
assortment, preferential 105, 203
  random 105, 138
attractiveness 126–31
attractor, strange 54
automaton 28
  cellular 13–14, 22–4
  kinematic 20–2
  self-reproducing 17–26, 28
average 54, 73, 76–8, 168–9, 187
Axelrod, R. 185, 187–90, 192, 198, 201, 203–4

baboon 175, 202
bacteria 68, 70, 72, 101, 109, 203
Bakker, T. 128
balance, shifting 87–8
Bartz, S. 116
bee 102, 118–19, 139, 176
Bell, G. 148
Bergerud, A. T. 61
bet 76, 78, 80
bias 171
bifurcation 52
  cascade of 52, 55
bird 119, 155, 159, 171
bistability 57, 61, 170

bluff 160–2, 174
Boltzmann, L. 212
bottle experiments 48, 57
Boyd, R. 195
branching process 67–71, 78
breeding 97, 120, 135
Brown, G. W. 169
Burks, A. 23
butterfly 108, 174
  effect 53–4, 65
Byl, J. 24

Cairns-Smith, A. G. 24–5
card games 7, 74, 168, *see also* poker
caribou 46–7, 62
cat 207–8
caterpillar 119
chain reaction 68–9
chaos 50–5, 61, 63, 65, 170, 196
cheater 174, 210
chess 86, 120, 159–61, 173, 186, 203, 213
'chicken' game 160, 163–5, 170, 172–4
chimpanzee 202
chromosome 93, 102–4, 109–11, 121, 140, 142, 145–6
ciliate 128, 141
cleaner fish 204
clique 131, 189
clock, molecular 89–90, 92
Codd, E. F. 23, 27–8
coevolution 42, 148–53
coexistence 45, 56–7, 170
Colinvaux, P. 46
colour, of eyes 96–7, 105, 121
  of skin 105
  warning 119, 194
colonization 63–4
communication 174
community 63–4, 201–5
competition 49, 56–9, 61
  local 168
  male 132
  sib 147–8
complexity 65, 121–2